DISCARD

ELECTRONICS
PROJECTS FOR YOUNG SCIENTISTS

GEORGE DeLUCENAY LEON

ELECTRONICS PROJECTS FOR YOUNG SCIENTISTS

FRANKLIN WATTS
NEW YORK / LONDON / TORONTO / SYDNEY
1991

Illustrations by Vantage Art

Photographs courtesy of the author

Library of Congress Cataloging-in-Publication Data
Leon, George deLucenay.
Electronics projects for young scientists / George deLucenay Leon.
p.   cm. — (Projects for young scientists)
Includes bibliographical references and index.
Summary: Instructions for a variety of projects illustrating the
principles of electronics.
ISBN 0-531-11071-0
1. Electronics—Experiments—Juvenile literature.
[1. Electronics—Experiments. 2. Experiments.] I. Title.
II. Series.
TK7820.L38   1991
621.381′078—dc20
91-17823 CIP AC

CONTENTS

ELECTRONICS
PROJECTS FOR YOUNG SCIENTISTS

PREFACE

There were two purposes in writing this book: (1) to give beginners some basic ideas about electronics and (2) to guide readers through a variety of useful, even amusing, projects that will increase their knowledge. It is hoped that the projects in this book will be suitable for classroom laboratory use and for science fair projects.

Each project is designed to be progressively more difficult than the previous one. If readers wish to, they may skip the first section and jump right into building whatever project captures their interest. Every attempt was made to choose projects that use components that are easily available to experimenters, no matter where they live.

There's fun in store!

INTRODUCTION

Electronics is defined as the branch of physics which deals with the emission, behavior, and effects of electrons. An electron, a tiny part of an atom, is an elementary negative particle. The electrons that are part of an atom's structure are held in place normally by means of weak forces of attraction from the core of the atom. By applying a pressure—such as one generated by a battery—these forces are overcome and electrons are made to flow along a metal path. Typically a wire forms such a path. A current is created with the electron flow from the negative terminal of the source of pressure (the battery) to the positive terminal.

Think of electrons as being similar to tiny drops of water, but so tiny that they are invisible. Many drops become a large amount of water. This water can then be made to flow through a pipe, just as electrons flow through a metal wire. All of those invisible electrons become a large flow of negative electricity. No matter how much electricity is flowing along a wire, we can't see it. Transmission lines carry millions of volts, but the electricity remains invisible. We see the result when electricity is used to turn on a light bulb, for example. And we certainly can feel it if we make the mistake of touching a live wire!

As you work your way through the projects in this book, you will usually be dealing with very small voltages to create a variety of interesting experiments. This includes switching devices, sound amplifiers, test instruments, and many other fascinating projects. However, you will find a few projects that require the use of house voltage, which is much larger. **Remember that such electricity is not only dangerous; if you become careless it can be *fatal*. So be extremely careful whenever you use house voltage. You should do those experiments that use house voltage, such as when you are building a power supply (Chapter Five), *only under adult supervision.***

You are about to enter the door to the world of a science that now controls everything from radios to computers to space shuttles. Electronics has enabled us to design complex robots that can work in areas too dangerous for human beings. Such robots can handle dangerous radioactive wastes, for example.

THE BEGINNING OF THE ELECTRONIC AGE

On December 23, 1947, three scientists from Bell Labs in New Jersey—William Shockley, Walter Brattain, and John Bardeen—found that by placing two wires on a germanium crystal and attaching a low voltage, a voice over the telephone was amplified forty times. The three men had invented the transistor, and on that day, electronics came of age.

But electronics began long, long before that, almost 3,000 years earlier. The credit for its birth belongs to Thales, a Greek mathematician and philosopher who lived on the west coast of Asia Minor from 640 to 546 B.C. He was the first individual we know of who discovered the presence of static electricity and magnetism. He is also credited with coining the word *atom,* meaning the very smallest part of matter. Of course, we know now that atoms are made up of still smaller particles, and that these particles are composed of smaller particles, and so on.

By rubbing certain materials—amber is one of many—Thales found that they attracted feathers and other lightweight objects. We call such materials *insulators* because they do not conduct electricity. A typical insulator is the rubber wrapped around electric wire to prevent the electricity in the wire from escaping to other metal objects. In varying degrees most metals are *conductors* because electric currents flow easily through them. Silver is the best conductor, and aluminum the worst.

The true beginning of the science of electronics was the invention of the vacuum tube in 1874. Improvements were made continually over the years. The first tube had two elements and was called a diode. Subsequently a three-element tube, the triode, was invented. Then along came the pentode, with five elements. Tubes improved in many ways, but they remained large, required high voltages, and, because of the heat they generated, had a fairly short life. This was measured in a few hundred hours under ideal conditions. Transistors, on the other hand, are tiny, require low voltages, and seem to last forever.

One example dramatically illustrates the difference between vacuum tubes and transistors. During World War II, before the transistor was invented, an early digital computer using vacuum tubes was built. It filled a very large room because of the thousands of tubes in its circuitry. Because of their short life dozens of tubes had to be replaced every day. Compare that with today's lap-top computers, which weigh as little as 4 pounds (1.8 kilograms [kg]) and have a memory that puts the old tube-driven computers to shame.

Then microelectronics was born when the microchip was invented by Robert N. Noyce. One microchip, half the length of your little finger, contains an entire circuit: numerous transistors, resistors, and capacitors etched together on a thin film of metal only a few thousandths of an inch thick. Because there are no wires connecting the individual components, there are no variances in behavior and no "noise" (unwanted signals). Interconnecting all of those components within a single chip saves space and time in soldering.

If it weren't for these *integrated circuits,* as microchips are called, space missions to the ends of our solar system would be just a dream. And of course, without the chips we could not have tiny radios, 2-inch (5 cm) TV sets, and a host of other miniature devices. Computers, videocassette recorders (VCRs), washing machines with their intricate cycles, and car and home alarms would be huge, cumbersome affairs. A "memory" chip built into a chessplaying computer can perform 800,000 analyses of chess positions in 1 second! And this is only the beginning.

This book will show you how to build projects that illustrate the principles of electronics. At the same time you will acquire some handy gadgets. You will begin by using transistors; then you will experiment with more complex circuits that employ different types of integrated circuits.

1

OHM'S LAW

As water flows from a higher level to a lower one, so electricity flows from the negative pole to the positive pole. But for the sake of analyzing circuits, we treat the flow of electric current as though it went from positive to negative. It seems easier (for some reason) to think of it that way.

Let's set up an experiment that demonstrates that if you place an electrical conductor, a wire for example, between two oppositely charged bodies—one body having more electrons that the other—the electrons will move until both bodies are equally charged. You can see how this fits in with the electronic theory defined earlier. Think of a circuit consisting of a battery connected to a lamp. Once the circuit is completed, free electrons flow from the negative (minus) side of the battery through the lamp to its positive (plus) side. These electrons are produced by the chemical action within the battery. As they flow, they furnish current to the lamp, causing it to light. This proves that a current will flow between two oppositely charged bodies—the plus and the minus sides of the battery.

VOLTAGE

To cause a flow of free electrons, there must be an electric charge differential between the ends of an electric conductor. One end of the circuit must have a different quantity of electrons from the other to allow the current to flow. It's almost like saying that water will flow only downhill. These flowing electrons form the electric current. The difference between the two ends of the conductor pushes the electrons through the conductor. This pressure is called *electromotive force (EMF)* or *voltage.* The EMF is measured in *volts* and is represented by the letter *V.*

CURRENT

The greater the charge or difference between the ends of the conductor, the greater the pressure in the circuit. This pressure (current) is measured in *amperes.* Its symbol is the letter *I.*

RESISTANCE

A conductor offers some resistance to the flow of electrons. Some metals are more resistive to the current flow than others. Silver offers the least resistance. Gold, copper, and aluminum offer more and more resistance, in that order. Some metals do not conduct at all. Many electronic components, such as transistors and diodes, to name only two, conduct in only one direction. Semiconductors are so called because they conduct more than insulators but less than true conductors. In other words, they are between conductors and nonconductors.

Resistance, symbolized Ω (for omega, the last letter of the Greek alphabet), is measured in *ohms.*

When an electric current flows through resistance, heat is generated. The greater the current and the greater the resistance, the greater the amount of heat. Although normally resistance is not helpful in a circuit, it does come in handy in some cases. A toaster or any heating element works because its wires offer so much resistance that they produce much heat.

VOLTS/AMPERES/RESISTANCE

To explain the relationships among volts, amperes, and resistance, let's return to the example of water flowing through a pipe. The amount of water coming out at the end of the pipe can be likened to the amount of electricity, or voltage, in a circuit. If a pump is attached somewhere along the pipe to push the water along, more pressure is exerted. Then we have the equivalent of greater current, or more amperes, in the circuit. But if we have something in the pipe that reduces the flow—a narrowing of the pipe, for example—then we have resistance (R).

OHM'S LAW

There are very definite relationships among voltage, current, and resistance. If you have an ammeter (or a volt ohm meter [VOM]), a few resistors, two batteries, and some wire you can set up an experiment that will show these relationships. Figure 1-1 illustrates the relationship (Ohm's law).

The various forms of Ohm's law are the most important facts you must know and *understand* to learn about electronics.

Let's take a look at Figure 1-2. In !-2A we have a very powerful 9-volt (V) battery—or any source of current—and a 1-ohm (Ω) resistor. The meter indicates that 9 amperes (A) is passing through the circuit. $I = V/R$ or $I = 9$ V/1 $\Omega = 9$ A.

But if we keep the same resistance—1 Ω and double the number of batteries—as in Figure 1-2A—the meter will show that 18 A is flowing through the circuit. Then we have $I = 18$ V/1 $\Omega = 18$ A.

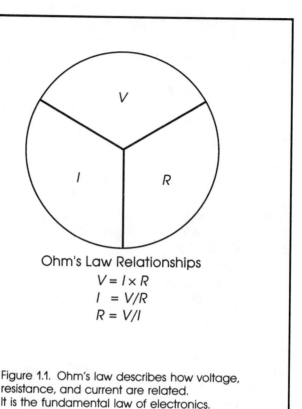

Ohm's Law Relationships
$$V = I \times R$$
$$I = V/R$$
$$R = V/I$$

Figure 1.1. Ohm's law describes how voltage,
resistance, and current are related.
It is the fundamental law of electronics.

In Figure 1-2C we doubled the resistance: 2Ω instead
of $1\,\Omega$. We used the same 9-V battery, but now the meter
shows that only 4.5 A is flowing. $I = 9$ V/2 Ω, or 4.5 A.

You have now seen the relationships among the three
terms, as it is stated in Ohm's law. Your meter may not
show exactly the same numbers. This may be due to the
batteries' voltage, or resistance may have been greater
than $1\,\Omega$. The preceding examples are meant to show the
relationships of the various elements expressed in Ohm's
law. This law should be memorized and never, never for-
gotten.

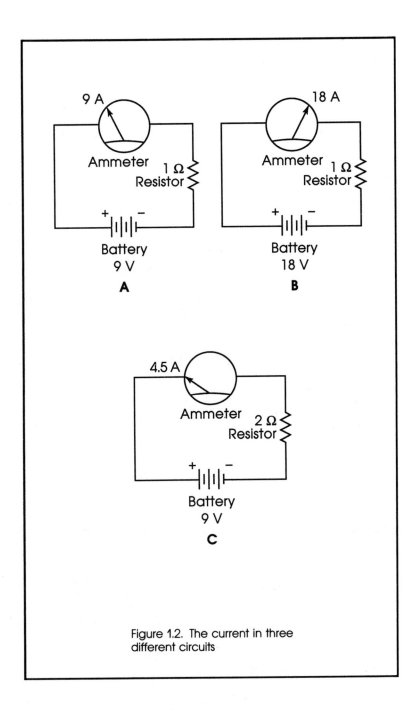

Figure 1.2. The current in three different circuits

We found in our experiment that

$$I = V/R$$

where I = current, V = voltage, and R = resistance. This form of the equation gives us the value for *current*.

We use a different form to solve for *voltage* by writing the equation in this fashion:

$$V = I \times R$$

And finally to get *resistance* we turn the equation around:

$$R = V/I$$

From Ohm's law we can derive still another important fact. We can find the amount of power in a circuit: power in watts and expressed by the letter P. The formula is

$$P = V \times I$$

Figure 1-3 shows the relationships of V, I, and R in a power equation. This is illustrated by Figure 1-2C, where we found that we have 9 V and 4.5 A. Using those facts to substitute for that last equation we have

$$P = 9 \text{ V} \times 4.5 \text{ A} \quad \text{or} \quad 40.5 \text{ W.}$$

In most projects you will be more likely to encounter fractions of a watt. These are expressed as *milliwatts* (mW), for thousandths of a watt; *microwatts* (μW), for millionths of a watt; *millivolts* (mV) for thousandths of a volt; and microvolts (μV) for millionths of a volt. Table 1 shows the values of volts, ohms, amperes and watts in numerical form.

POWER RELATIONSHIPS

By employing the power formula and applying it with Ohm's law, we can derive a number of very valuable power formulas that can be used with direct current circuits. As you know, $V = I \times R$. You also know that $P = V \times I$.

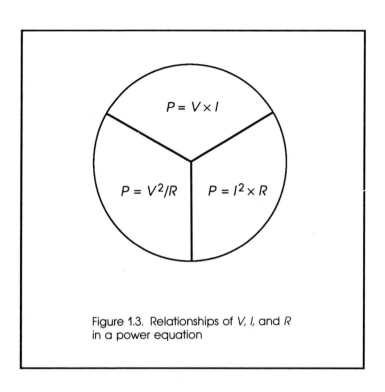

Figure 1.3. Relationships of V, I, and R
in a power equation

TABLE 1
COMMON MEASUREMENTS USED IN ELECTRONICS

Unit	Symbol	Multiple	
volt	v	kilovolt (kv)	= 1,000 volts
volt	v	millivolt (mv)	= 1/1,000 volts
ohm	Ω	kilohm (KΩ)	= 1,000 ohms
ohm	Ω	megohm (MΩ)	= 1,000,000 ohms
ampere	A	milliampere (ma)	= 1/1,000 ampere
ampere	A	microampere (μa)	= 1/1,000,000 ampere

Substituting $I \times R$ for V in the power formula you now have

$$P = I \times R \times I \quad \text{or} \quad P = I^2 \times R$$

We can go further by substituting V/R for I in the formuia:

$$P = V \times I \quad \text{or} \quad P = V \times V/R$$

This now gives us $P = V^2/R$.

By means of substitution you can arrive at still other formulas. Try your hand at employing different voltages and different resistances. This will teach you how to calculate values within circuits. Being able to do that you can then check the values that really exist in your project after it is finished. If there is a great difference, you may have made an error in wiring or in choosing components. Never forget, however, that the theoretical values you obtain from applying Ohm's law may differ slightly from actual results, as has been pointed out earlier. Still, the difference should not be very great unless the circuit is at fault.

What this all means is that you should become thoroughly familiar with Ohm's law. Unless you know it backward and forward, you will never really be able to experiment with electronic projects.

ALTERNATING CURRENT

Most of the work you will be doing while building projects in this book involves *direct current* (DC): an electric current that flows in only one direction. However, there is also *alternating current* or *voltage* (AC): an electric current that is constantly reversing its direction at regular intervals. It starts from 0 V, flows in one direction, rises to a maximum value, then reverses itself, going once more from zero to maximum. This is called one *cycle*.

Our alternating current—that in ordinary house current—changes direction sixty times per second. A cycle per second is called a *hertz* (Hz). A complete wavelength is represented by the Greek letter *lambda* (λ). So we say that our house current flows at the rate of 60 Hz, or 60 cycles per second. Our clocks and many motors depend on this regular, constant change in current direction to maintain their regularity.

When you start building projects you will notice that although you use AC (from a wall outlet), the alternating voltage is converted to DC by means of the components that make up the power supply. As you know, dry cells produce DC only. For most of the projects in this book DC will be used. Nevertheless, no theory of electronics is complete without a basic understanding of AC.

2

TOOLS AND COMPONENTS

Before you can start any experiments, there are a number of tools that you must have, plus some that are not absolutely necessary but will make your work easier.

Tools you must have:

Low-powered soldering iron—15 to 25 W
Roll of rosin core solder
Needle nose pliers
Screwdriver
Tweezers
Wire cutter
Roll of no. 24 single-strand insulated wire
Roll of no. 20 single-strand insulated wire
Hand or electric drill and assorted bits
Files to smooth holes
Roll of insulating tape
Breadboard

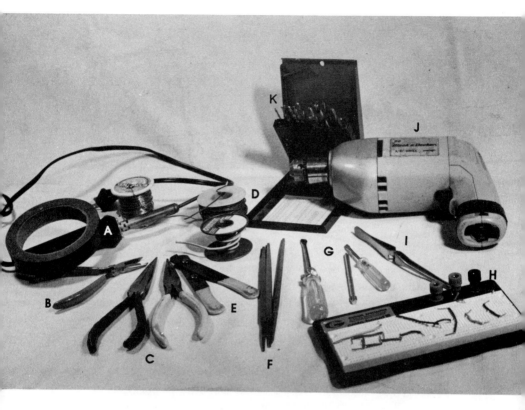

**A basic set of tools for
electronics experimenting:**
(A) soldering iron
(B) wire cutter
(C) pliers
(D) rolls of wire
(E) wire stripper
(F) files
(G) screwdrivers
(H) breadboard
(I) tweezers
(J) drill
(K) bits for drill

Some tools that are nice to have:

Wire stripper
Volt-ohm-milliammeter (almost a necessity)
Magnifying glass
Bench-mounted vise

Why a low-wattage soldering iron? Semiconductors such as transistors, diodes, and integrated circuits (ICs) are sensitive to heat. The rule is to apply low heat only long enough (about 10 seconds) to make a solid soldering joint. First heat the iron and melt solder over its tip until it is entirely covered.

Most nonworking projects result from improperly soldered joints. Twist the wires so that you have a good physical joint. Now apply the hot iron to the joint of the two wires. Bring the solder to the point until the solder melts and flows over the connection. Remove the iron and hold the joint still for a second or two. The solder on the joint should be glossy smooth. If it isn't, then do it again.

A rough-looking solder joint is one that was not completely heated. It is bound to be a source of trouble. Prevent it by soldering carefully. Practice on a few scraps of wire until you feel confident you can make a good electrically sound joint. Some portable projects may be shaken, so you must make certain that this effect will not loosen the joint.

To prevent the heat from damaging a semiconductor, hold the wire to the semiconductor with the needle nose plier or use a heat sink. A *heat sink* is a small clamp that prevents the heat from traveling up the wire and damaging the semiconductor. Either must clamp the wire between the solder joint and the semiconductor.

Then examine the joint with a magnifier. Does the joint look solid? Make certain that it is.

As soon as you can, buy a VOM. It measures resistance, volts, and current. It allows you to make certain that

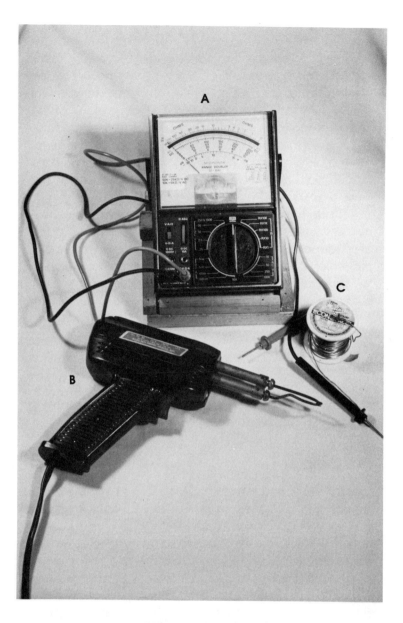

**Other tools that will
make your work easier
and faster: (A) volt-ohm meter;
(B) solder gun; (C) solder.**

a connection has been made correctly, and it will also tell you how much voltage a battery is producing. It will save you money, as you will not have to buy several different meters. In almost every circuit you will find many of the same component parts. Some—such as resistors and capacitors—are so basic that you will run into them in practically every circuit. Not all the circuits have the same parts, but the parts described here are the ones you are most likely to encounter.

The term *solid-state* is used so often that it must be defined. *Solid-state* describes a component whose operation depends on the control of electric or magnetic properties in solids. These include transistors and diodes, to name only two such devices. Such components differ from vacuum or gaseous devices.

BATTERY

A *battery* is a source of DC voltage that consists of two or more cells. These cells convert chemical, solar, or some other form of energy into electrical energy. The symbol for a battery is $+\ \underset{|\,|\,|}{\quad}\ -$

CAPACITOR

A *capacitor* is a device that is made up of two conducting surfaces separated by an insulating material such as air, paper, oil, or mica. Originally it was called a condenser because it stores electrical energy. It holds electrical energy for a while, blocks the flow of direct current, and permits the flow of alternating current. The amount of AC that it passes is determined by the capacitance of the unit and its frequency. Capacitance is measured in *farads* (F). However, this is such a large amount that the *microfarad* (one-millionth of a farad [μF]) is used more commonly. Still smaller units are micromicrofarads, now referred to as picofarads (pF). Consult an electronics manual for various color codes found on capacitors in surplus equipment. The codes indicate capacitance.

Correct polarity must be observed but only when connecting an electrolytic capacitor. This is a long tubular unit. The plus end is marked (+) and must be wired as indicated in the schematic. Its value is stamped on the unit.

If a circuit calls for a capacitor with a value of, say, 500 μF at 25 V, you can substitute a unit with a higher voltage—never lower. You can also, if you must, use one with a higher capacitance. But always try to get the required capacitance if possible.

Disk capacitors have a much lower capacitance and have no polarity. They can be connected in either direction. Their value is marked on the unit. Capacitors also are available in many shapes and sizes. If the unit required is an electrolytic, the plus (+) and minus (−) will be indicated in the schematic. The symbol →)⊢ or ⌗⊣⊢ represents a capacitor.

You can combine capacitors in series or in parallel to provide a value that is different from that of the individual capacitor. As you can see in Figure 2-1A, the capacitance of capacitors in parallel is equal to the sum of the individual capacitors.

The total capacitance of capacitors in series is always less than that of the smallest capacitor. In Figure 2-1B the formula for two capacitors in series is shown; Figure 2-1C shows the value for the number of capacitors in series. Any number of capacitors (above two) can be used in this equation. By applying the equations, you can combine two or more units to provide the exact capacitance you need but cannot obtain in one unit.

CHOKE

A *choke* presents inductance to an AC circuit and is used to offer high impedance to frequencies above a speci-

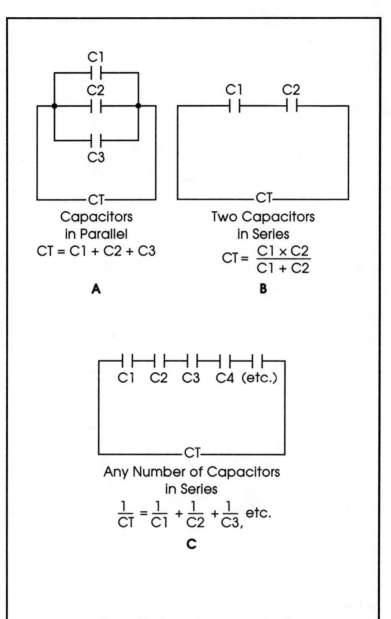

Figure 2.1. Capacitance equations

fied frequency range without affecting the flow of DC. Inductance is measured by a unit called a *henry* (H); the *millihenry* (mH) and the *microhenry* (μH) measure smaller quantities. The symbol ⌐ represents a choke.

DIODE

A *diode* is a two-electrode semiconductor that rectifies the current (causes it to flow in only one direction). In power supplies the diode acts as a rectifier to convert alternating current to pulsating direct current. There are several types of diodes. Regardless of the type, they all have a polarity that must be observed, as they pass current in only one direction. Observe the polarity when wiring diodes into a circuit. The rectifier diode has a band across one end. This indicates the *cathode* or negative side. The other end—the *anode*—is the positive side. (This is discussed further in the sections on light-emitting diodes (LEDs) and zeners.) The symbol ▶◀ represents a diode.

FERRITE ROD
LOOPSTICK ANTENNA

The *ferrite rod loopstick antenna* is a type of antenna used instead of a loop antenna. It consists of a loop wound around a ferrite rod and is seen commonly on car and pocket radios. Its symbol is

FUSE

A fuse is a protective device usually consisting of a piece of thin wire that melts and breaks the circuit when the current exceeds a specific value. Fuses are rated in amperes. A 1-A fuse will flow if more than 1 A is flowing through the circuit. For a circuit in which 5 A will flow, you must have a fuse rated slightly above that figure. The symbol represents a fuse.

GROUND

Although there may not be an actual connection to the earth, a circuit is said to be at *ground* potential at that point. This means that if it were connected to the earth, it would not disturb the operation of the circuit. In some cases, all ground connections are made to a common point or to a *ground bus:* a wire to which ground connections are made. This bus is then connected to the negative output. In others, the ground connections are to the metal chassis and joined at one point close to the output.

The symbols vary as indicated:

Common ground: ⏚   Chassis ground: ⏚

INTEGRATED CIRCUIT

An *integrated circuit* (IC) is an electronic device in which a number of active and passive components are combined in one package. Symbols vary widely, depending on the type of IC and its function. You may notice on some schematics that an IC number, such as SN7413, may be followed by a letter such as *A* or *N*. The units are basically the same. The letter signifies an industry or military standard. As an experimenter you can ignore these letters.

LIGHT-EMITTING DIODE

A *light-emitting diode* (LED) is a two-element device that lights up when biased in a forward direction. The symbol represents an LED.

METER

A *meter* is one of a variety of electromechanical devices used to measure current, voltage, resistance, and volume levels. The symbol represents a meter.

PHOTOTRANSISTOR

A *phototransistor* is a type of transistor whose base is exposed to light through a lens in the housing. As the light increases, so does the collector current. This is the result of the amplification of the base current. (See the discussion of the transistor.) The symbol or

indicates a phototransistor.

POTENTIOMETER

A *potentiometer* is a device used to divide resistance by means of terminals connected to both ends of a resistance element and a third terminal connected to the wiper contact. The voltage is reduced as the wiper moves over the resistive element. This makes it possible to change the resistance mechanically. The term is also used for a variable resistor. A "pot"—as it is sometimes called—is used also as a volume control.

The unit of measurement is an ohm. (See the discussion of the resistor.) The symbol for a potentiometer is or

RELAY

A *relay* is an electromechanical device whose contacts are opened or closed by switching a control voltage on or off. Depending on the position of these contacts, a second set of contacts connected to another circuit is also opened or closed. Varying the conditions in one circuit changes the conditions of another part of the circuit or of a different circuit. The symbol represents a relay.

RESISTOR

A *resistor* is a device that offers resistance in a circuit. There are two basic types of resistors: fixed and variable. The resistor is rated in two ways: the first is the amount of resistance that it will offer to the circuit; the second is the amount of wattage it can handle. The smallest is 1/4 W; the largest is several watts. The symbol represents a resistor.

**Resistors can be large or small.
The two round types shown are variable types;
the four long types offer fixed values.**

The ordinary fixed carbon resistors have three and sometimes four colored bands. The first three bands indicate the value of the resistor. The fourth band indicates the tolerance of that unit. (Consult an electronics manual to identify transistors.)

Most circuits are not critical, so a three-band resistor, which has a 20 percent tolerance, is adequate. (These resistors are not offered commercially, but you will run across them when you buy surplus equipment.) This means that a 1K resistor—1,000 Ω—may actually have the value of 800 to 1,200 Ω. Its color bands are—reading from one end— brown (1), black (0), and red (100).

If you have a VOM, you can measure the resistance of a specific unit. Few circuits are so critical that they demand the exact resistance. But a resistor may be mismarked by the manufacturer, and it is a good idea to check its value to make certain it will not upset the circuit.

In few projects do you need a resistor with a lower resistance than one marked with a silver band—10 percent tolerance—or, much less, with a gold band, which indicates a 5 percent tolerance. They are more expensive and are needed only in the most critical circuits.

There are also military units that have only 1 percent tolerance. These are still more expensive, but you can find them occasionally among surplus equipment. When you do, you buy an entire board that contains enough parts to make it worth your while to unsolder the parts you need. This allows you to fill your "junk" box, the resistors cost you very little, and usually you also find many other "goodies."

There is still another way of getting a specific resistance for a DC circuit when that value is not available. Connect two resistors in parallel or in series. See Figure 2-2. There you will notice that the equations for resistors are similar to those for capacitors except that the equation for two resistors in parallel is like that of the equation for capacitors in series. What it means is that the more resistors you have in series in a DC circuit, the greater is the total

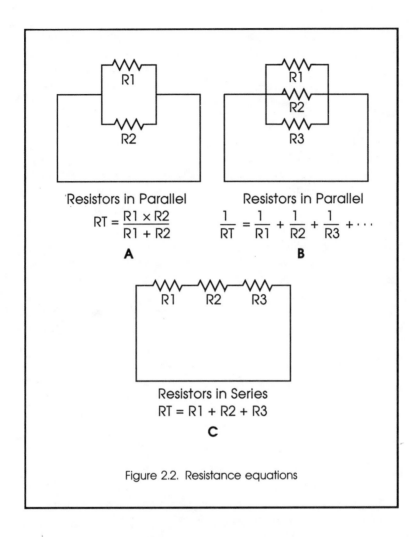

Resistors in Parallel

$$RT = \frac{R1 \times R2}{R1 + R2}$$

A

Resistors in Parallel

$$\frac{1}{RT} = \frac{1}{R1} + \frac{1}{R2} + \frac{1}{R3} + \cdots$$

B

Resistors in Series

$$RT = R1 + R2 + R3$$

C

Figure 2.2. Resistance equations

resistance, whereas capacitors in series reduce the total capacitance.

In Figure 2-2A is the formula for two resistors in parallel. Total resistance will always be less than either R1 or R2. By substituting two resistors in parallel you can obtain a value that you cannot get otherwise.

Figure 2-2B shows three resistors in parallel. Apply the

formula in Figure 2-2A for two resistors and use the result to repeat the formula with the third.

Two or more resistors in series is the sum of all the resistors, as is shown in Figure 2-2C.

SWITCHES

Some switches are simple: they turn on a circuit and they turn it off. However, because of the variety of switch arrangements, as you can see in Figure 2-3, they can do much more complicated things. They allow you to com-

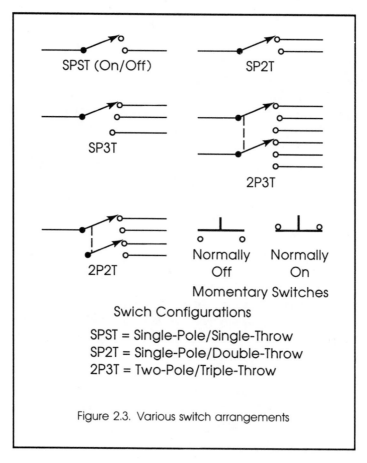

Figure 2.3. Various switch arrangements

bine various parts of a circuit—for instance, add more resistors or more capacitors to a circuit—without doing any
rewiring. Some switches may have two or even more plates
to which are wired ten or twenty components. The main
switch on a VOM is typical of a switch that combines many
components and many functions, all without needing
several additional switches or added wiring. The symbol _____/° _____ represents a switch.

TRANSFORMER

A *transformer* is an electrical device consisting of two or
more coils. Coupled together only by magnetic induction,
a transformer moves electric energy from one circuit to
another circuit while converting current or voltage to another value. What this means is that although the voltage
and current may change, the frequency will not, although
there is no physical connection between the coils of the
transformer.

A voltage flowing through one coil causes voltage to
flow in the other although they are not connected physically.

The ratio of the turns in a secondary coil to the turns
in a primary coil is known as the *turns ratio.*

If the turns in the primary are equal to the turns in the
secondary, then the transformer will produce exactly the
same amount of voltage and current as existed in the primary. A step-up transformer has more turns in the secondary than in the primary. You pay for that increased
(stepped-up) voltage by a corresponding decrease in
current. A step-down transformer, as may be expected,
has more turns in the primary than the secondary.

The wires to the primary side are usually black. Those
to the secondary may be green. If the secondary side is
tapped, then the third wire from that side is green and
yellow.

Audio transformers are used, as the name suggests, in audio circuits to match one part of the circuit which has a high *impedance* (a form of resistance in an AC circuit) to one which has a low impedance.

A transformer is rated according to the amount of voltage it can handle as well as the amperage. Some transformers are huge, weighing several pounds. Others weigh only a few ounces. But the symbol, as is the case for other components, is always the same, regardless of size or capability:

TRANSISTORS

A *transistor* is a semiconductive device made of silicon or germanium, containing three or more electrodes (see the symbol). The word comes from the combination of two words: *TRANSfer* and *resISTOR*. A transistor has numerous applications, including amplification and switching.

The most commonly used transistors are PNP and NPN. An *NPN transistor* is one that is made up of a P-type base sandwiched between N-type collector and emitter. A *PNP transistor* consists of an N-type base surrounded by P-type collector and P-type emitter. The three regions are illustrated in Figure 2-4. Both have three leads: base, collector, and emitter. Conduction is accomplished by electrons (negatively charged particles) and holes. These holes exist as mobile electron vacancies similar to positive charges.

The PNP type must have a *negative* voltage applied to the base and to the collector in order to conduct. The NPN type requires a *positive* voltage at the base and the collector. The direction of the arrow in the symbol tells you whether it is an NPN or a PNP.

The transistor in a project you will soon see, the crystal radio amplifier, acts as an amplifier. It takes a weak signal and amplifies (enlarges) it by a factor that may be large or small, depending on the circuit that surrounds the transistor.

**Transistors come in a variety
of shapes and sizes.**

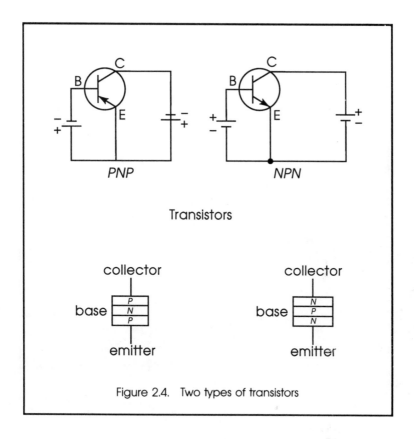

Figure 2.4. Two types of transistors

Transistor circuits are usually set up so that the base and the emitter form the input circuit. The collector and the emitter form the output circuit. You can see these circuits illustrated in Figure 2-5.

Transistors are manufactured in a variety of shapes. They do not look alike, but the transistor symbol usually remains the same. A typical symbol for a PNP transistor is

PNP   NPN

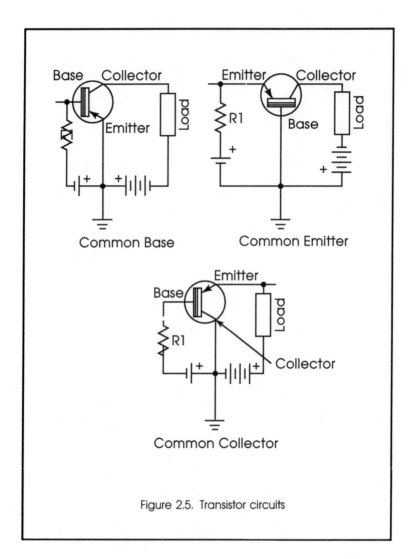

Figure 2.5. Transistor circuits

TRIAC

The *triac* is a semiconductor switch controlled by a gate and designed for AC power control. By controlling the gate signal load, current can be varied from 5 to 95 percent of full power. The triac symbol is

anode 2
anode 1
Gate

ZENER DIODE

A *zener diode* is a two-layer device that presents a sudden sharp rise in current when a specific reverse voltage (zener value) occurs. The *zener value* is that voltage at a specific current at which there is a change in characteristics of the diode from those at a lower voltage. The symbol represents a zener.

One further point: In the circuits that follow, you will see a transistor, diode, or IC listed with a particular parts number. That does not mean that it is the only one that can be used. Radio Shack, Sylvania, Calectro, RCA, and GE, as well as several other manufacturers, have their own parts number for the same device. An electronic store will have cross-reference charts for the parts it handles.

3

READING AND WIRING
A SCHEMATIC

You want to get started in building a project, but you don't know exactly where to start. Look at Figure 3-1A. (This is only an example, not a working circuit.) It's confusing, at first, because the symbols don't look at all like the parts that are going to make up the unit.

There's a reason for the symbols. Not all resistors look alike, nor do all capacitors or many other electronic parts. To draw a picture of each would be complicated, so one symbol is used for all capacitors. The same holds true for the many types of resistors: one symbol for every type. The parts list that accompanies every schematic tells you exactly the type and value of each component. How do you follow the schematic and assemble the parts the way they should be?

BREADBOARDING

What's a breadboard? As the name suggests, many years ago experimenters borrowed the wooden board on which

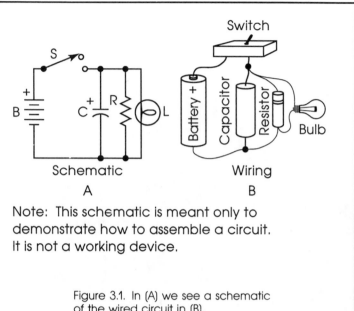

Switch

S

+
B

C +
R
L

Battery +

Capacitor

Resistor

Bulb

Schematic

A

Wiring

B

Note: This schematic is meant only to demonstrate how to assemble a circuit. It is not a working device.

Figure 3.1. In (A) we see a schematic of the wired circuit in (B).

bread was cut. They drilled holes or drove nails into it. Then they placed the leads of the components in the holes or wrapped them around the nails. Nowadays you do not have to sneak the breadboard out of the kitchen for your experiments. Buy a commercially made board with all the holes drilled. These boards also have internal wiring to make the connections even easier. They are made in various sizes and can be found in most electronics stores.

Figure 3-1B shows how the circuit in Figure 3-1A should be assembled. As you can see, it follows the schematic pretty closely. Just for practice get out your breadboard and follow the instructions.

You connect one side of the switch to the positive side of the battery. The easiest way is to place the battery in a holder and solder one wire to the holder and the other end to one pin of the switch.

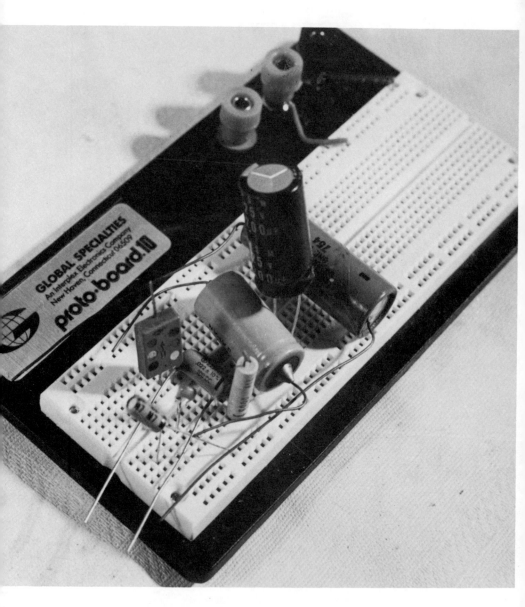

**Capacitors can be breadboarded
(as seen above), or wired either
horizontally or vertically.**

To the other pin of the switch, connect the positive end of the capacitor, one end of the resistor, and one side of the light bulb holder.

The wire from the negative end of the battery is connected to the negative end of the capacitor, to one end of the resistor, and to the other side of the light. Three connections are made at that point.

Not all circuits are soldered together (or *hard wired*). A breadboard is useful when you want to test the circuit and don't want to make it permanent by soldering the parts together. If you want to make a project you plan to keep, there are six ways to make a permanent circuit:

1. Hard wiring

2. Perforated board

3. Metallic tape

4. Etching process

5. Wire wrapping

6. Photographic process

HARD WIRING

Hard wiring is the oldest and simplest method. Following the schematic, you solder one lead of a component to another and solder wires between the components. This is fine when the circuit is simple, the components are large, and you do not need to save space.

PERFORATED BOARD

Perforated boards have holes already drilled and equally spaced from each other. You buy the board and the pins from an electronic store. The pins are made to fit exactly into these holes. The pins act as sockets for the component leads as well as for the wiring point. To save space, some components are mounted vertically. You can see that the layout is the same for the perforated board and the breadboard. This is why the breadboard is used to design and to test a circuit before it is made permanent.

METALLIC TAPE

Instead of using wires for the connections, you substitute strips of metallic tape that have glue on one side. You press down the tape onto a perforated board. You push the lead of a component through the hole in the perforated board and through the tape. Using a low-wattage iron you very quickly solder the leads of the component onto the tape. This process requires a great deal of care.

ETCHED BOARD

From your local electronic store buy a copper-clad perforated board. At the same time buy a can of resist—a type of varnish—and a fine brush. You will also need etching acid. *Extreme caution should be used with the acid.* Do not let it touch your skin or your clothes as it will burn the skin and make holes in the clothes. You can buy all the material as a kit.

One side of the board is covered with a coating of copper. Lay out the components on the board, following the schematic. This will give you an idea of how big a board you need, and where the components should be placed for the shortest lines between them. If two lines must cross each other, you "jump" over one line by bridging with wire so that it does not touch the line under it. That will be done later. Draw lines between the components with the brush dipped in the resist varnish.

Place the board in a plastic dish and cover it with acid. Use a wooden stick to stir the acid until the copper is dissolved. (Use caution with the acid.) When the copper has disappeared, remove the board and rinse it under running water. Remove the protective varnish with turpentine. Only the lines you drew should remain.

The copper lines are conductors. Drill holes to hold the leads. Push the leads of the components and bend them so they make good contact with the copper lines. Use a low-wattage iron and solder them in place. Or you can push pins through the holes and solder those in place instead. This type of construction is not as easy as the others, but it does provide a circuit that will stand up very well. It is especially well suited to the larger circuits.

WIRE WRAPPING

Save this technique for when you have several integrated circuits (ICs) on the same circuit. The ICs are fitted into special sockets that fit into a board with holes closely spaced. The sockets have tiny pins, and by using a wire-wrapping tool you wind fine wire from a pin to other points.

PHOTOGRAPHIC PROCESS

The photographic technique is used mostly when you want to make several copies of the same circuit. You must make one original and photocopy it to make as many copies as you need. This is a complex method and is best used when you have more experience and need several copies of the same board. Before you make copies be certain that the original is perfect.

MAKING THE MOCK-UP

You should prepare your electronics projects in a plastic or metal box, or *chassis*. But drilling holes all over a chassis without knowing in advance where the parts fit results in a project that looks like a piece of Swiss cheese. Once you've drilled holes in a metal or plastic chassis, the only way to

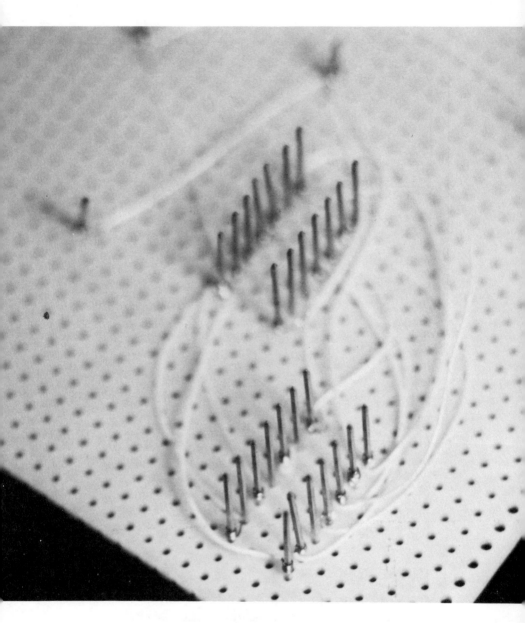

Wire-wrapping: Wire is wrapped around pins to interconnect a complicated project containing chips.

make the parts fit where they should, if you've made a mistake, is to drill more holes.

It's easy to prevent such mistakes with a little preparation. Make a cardboard *mock-up:* a copy of the chassis you plan to use.

Let's take as an example the project in Chapter 5 shown in Figure 5-1: a 0- to 15-V regulated power supply. You're not going to build it yet, but we're going to use it as an illustration of how to create a cardboard version of the chassis you would need to enclose the unit. The principle for making a mock-up is the same regardless of the size of chassis or the type of project you plan to use.

The first thing to do is assemble all the parts you will need. Let's say that when you have all the parts in front of you, you feel that you need a chassis about $6 \times 6 \times 2$ inches ($15 \times 15 \times 5$ cm). You look in a parts catalogue and find a chassis that measures $7 \times 5 \times 2$ inches. That's close enough. It's better to make it a little too large than too small.

Get a piece of cardboard about 9×11 inches. Using a pencil, mark it off as in Figure 3-2. With a sharp knife cut off the shaded areas. Now *score* along the dotted lines, being careful not to cut through the thickness of the board. Bend the board along the scored lines. Tape the corners together so that now you have a box measuring $7 \times 5 \times 2$ inches ($17.5 \times 12 \times 5$ cm). You don't need a cover for the mock-up.

If you were actually to build the project, you would then punch holes wherever necessary. The mockup allows you to correct any mistakes, such as a part that is too large to fit into the space you planned for it. Remember that you are dealing in three dimensions. Is there room below the variable resistor so that you can mount it without difficulty? If it doesn't fit, it's easy enough to punch a hole higher up. This is what you should do first; make all the necessary corrections before you tackle the actual chassis.

The few minutes spent on getting everything right on

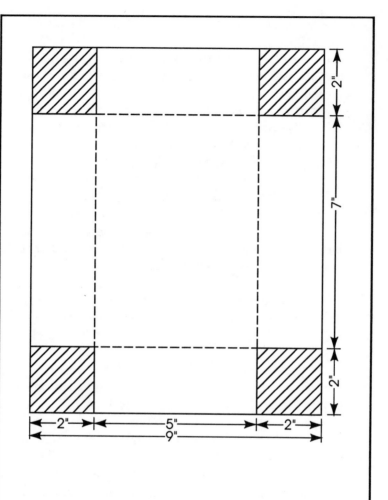

Figure 3.2. How to make a cardboard mock-up of the chassis for the 0-15 V regulated power supply described in chapter 5

the cardboard mock-up save hours of useless work and prevent you from drilling holes in the wrong places. Even the most experienced builder has made mistakes—at first. You save hours and create a professional-looking chassis if you follow these directions. The same principle applies regardless of the size of the project you have in mind.

Satisfied that everything is the way you want it, you will use the cardboard as a template for the holes to be drilled.

4

CONTINUITY TESTER, CRYSTAL RADIO, AMPLIFIED CRYSTAL RADIO, AND TELEGRAPH OSCILLATOR

BUILDING A
CONTINUITY TESTER

A continuity tester is a simple project to build, but that doesn't mean it has few uses. On the contrary, it is a test instrument that no one involved with electronics should overlook. Figure 4-1A shows the schematic and Figure 4-1B illustrates the actual wiring of the assembled device. The device is designed to show a break in a wire or loss of continuity between two points. It will also tell you whether the solder joint between two wires is well done. There must be *no current flowing* in the circuit when you employ the tester. Alligator clips can be used on the ends of the prods. The batteries inside the tester are connected so that the plus (+) end of one is touching the minus (−) end of the other. Use a battery holder that connects the two batteries. The *total voltage* is the sum of the voltages of the batteries. No switch is needed since the lamp is not drawing any voltage except when performing a test.

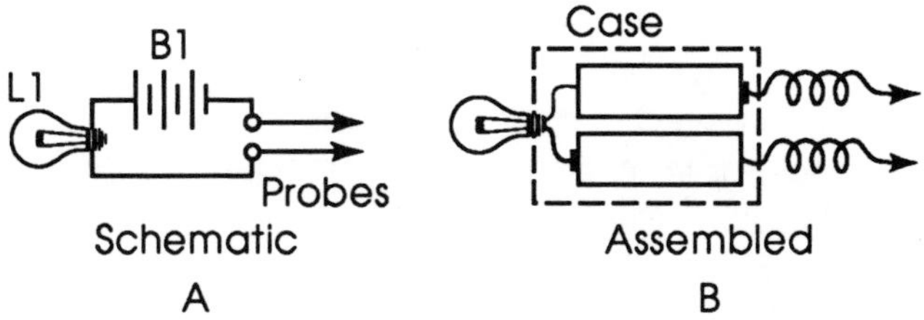

Figure 4.1. The schematic (A) and actual wiring (B) of the continuity tester

Parts List

B1: 2 1.5 V AA batteries

L1: 3 V light bulb

Case $2\frac{3}{4} \times 2\frac{1}{8} \times 1\frac{5}{8}$ inch

Battery holder

Light socket

Wire for probes

TESTING THE TESTER

Touch the two prods together. The lamp should light up. If not, check your wiring. Are the batteries making proper contact? Check the holders. If you still have a problem and you are certain the wiring is accurate, it is possible you have one or more poor batteries. This is another occasion when the VOM comes in handy. It must show a total of 3 V between the two ends of the batteries. Of course, it's much better to test the batteries *before* you place them in a circuit.

CRYSTAL RADIO

It's hard to capture the excitement of the first crystal radios. Imagine taking a crystal, "tickling" it with a tiny piece of wire connected to a pair of headphones, and actually getting sounds from the air. This is a modern version of the very first radio.

For this project we will use a breadboard. Note that when two lines cross like this ──┼── they are connected. They *are* *not* connected in these examples: ──┼── ──⟩── ──│──

Figure 4-2 is the schematic for the radio. In Figure 4-3 we have the components as they might be soldered to each other. All of the parts can be purchased in an electronic parts store or through a mail order firm.

The antenna should be as long as possible—10 feet (3 m) or so—and preferably as high up as you can get. An attic is a good place. Do not place it where it might touch an electrical wire and preferably do not locate it out of doors—especially if there is a possibility of lightning. Your ground can be a clamp to a cold water pipe.

C1, a variable capacitor, is your tuning device. Once your project is assembled, put on a set of high-impedance headphones. Do not use the type you get with shirt-pocket radios. These are of low impedance—usually 8 Ω—and will not work in this circuit. Now rotate C1 very slowly until you hear a station. It will be faint. If you do not hear any stations at first, move the antenna around.

Once you've tried the assembly on a breadboard, you will want to solder the parts together. Keep the wires between the components as short as possible.

AMPLIFIED CRYSTAL RADIO

You won't get ear-shattering volume from the crystal radio. You can, however, amplify the sound so that you can hear it through a loudspeaker with the next project (Figure 4-4).

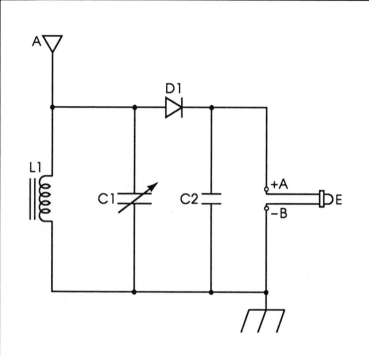

Parts List

Antenna: As long as possible
L1: Ferrite rod loopstick antenna
C1: 365 pF variable capacitor
C2: 0.001 µF disk capacitor
D1: 1n34A or similar germanium diode
E1: High-impedance earphone

Figure 4.2. Schematic of crystal radio

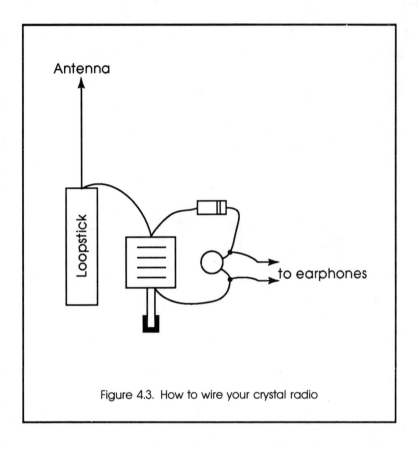

Figure 4.3. How to wire your crystal radio

Add this circuit to the previous one on the same bread-board if there is enough space.

The circuit is a simple one. The transistor (Q1) is a PNP type that amplifies the tiny sounds from the radio. The audio transformer (T1) reduces the high impedance of the circuit to one which matches the lower impedance of the speaker. When connecting the transformer, T1, make sure that the side connected to the speaker is the 8-Ω side. The latter can be any small 8-Ω speaker. The symbol /// means that this version of the ground connection is to be attached to the chassis. It is best if the three grounds are attached together and grounded on

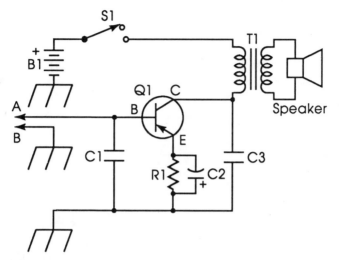

Parts List

B1: 9 V Battery
C1: 0.12 µF Mylar capacitor
C2: 15 µF, 10 V capacitor
C3: 0.15 µF disk
Q1: ECG 129 PNP transistor
R1: 150 Ω, $\frac{1}{2}$ W
S1: SPST switch
Speaker: 3 in, 8 Ω speaker
T1: Output cm transformer
     1 k Ω primary/8 Ω secondary

Figure 4.4. Schematic of crystal radio amplifier

only one point of the chassis, making this ground connection close to the output. Connect the amplifier to points A and B where the earphones were attached.

You should now hear any stations in your vicinity much louder. If you get too much interference from cars or from a nearby power station, wait until evening. You will probably have more luck then.

TELEGRAPH OSCILLATOR

Here's a project that will allow you to communicate with a friend using Morse code. The project is illustrated in Figure 4-5. Morse code is used by ships at sea as the audible dots and dashes that make up the code can be heard better than the human voice. This project converts the pressure on S1 to allow you to create long and short tones

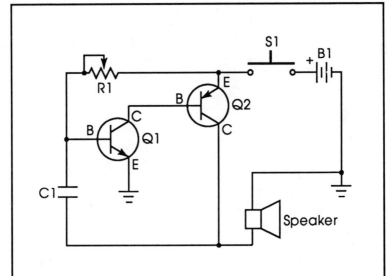

Parts List

B1: 1.5 V D cell
C1: 0.0022 µF capacitor
Q1: ECG123 NPN transistor
Q2: ECG128 PNP transistor
R1: 100K audio potentiometer
S1: Momentary on switch

Figure 4.5. Schematic of the telegraph oscillator

that can be heard by someone else. You can't send this over the airwaves, but you can run wires to another speaker or another set of headphones. If the person you want to communicate with lives far away, you can place your telephone next to your oscillator and your friend can leave his or her telephone off the hook. Your friend will be able to hear the tones you are sending and by using Morse code understand your message. If you build a second unit your friend will be able to send you a message.

Morse code consists of combinations of short dots and longer dashes. The letter *S* is made up of three dots. That means you press down S1—a momentary On switch—for just a fraction of a second. You do this three times. The letter *O* consists of three dashes. You hold down the switch a little longer three times.

R1 controls the volume of the sound that you will hear through the 8-Ω speaker. Limit yourself to a small speaker as you will not have enough volume to move the cone of a large speaker such as in a high-fidelity setup.

The two transistors—Q1 and Q2—amplify the sound while creating a tone each time the switch is depressed. The power is provided by a single B battery. Although this battery provides only 1.5 V, it will last a long time.

There are so few components that you can make it a portable unit. The size of the case depends on the size of the speaker. Drill a hole for the switch, and you are all set to send your coded messages.

5

POWER SUPPLIES AND THEIR
BASIC PRINCIPLES

Safety Note:
House voltage is not only dangerous but if you're care-less and make even *one mistake*, it can be lethal. Whenever using house voltage do it *under the supervision of an adult*. If you have never worked with electricity before, you should let an adult do the connections until you have learned how to make them safely.

BATTERY ELIMINATOR

We need electricity in some form to make most electronics projects work. This can come from the sun by means of solar cells or windmills, but more usually from an outlet in the wall—120 V—or from batteries such as A, AA, AAA, C, or D. Lantern batteries are also used when a long-lasting source of voltage is needed.

If the project is meant to be portable, batteries are the answer. For something stationary you generally use a power supply with a 120-V source.

It is true that a power supply is initially more expensive than batteries. Buying batteries every few days, however, costs more in the long run. Furthermore, when batteries become weak, they create "noise" within the circuit. If they remain unused within a device for a long time, they may leak and destroy everything they touch.

Battery-powered supplies are limited to low voltages and low current demand. If a project doesn't have to be portable and higher voltages or currents are needed, then the power must be supplied from the house current. This is about 120 V AC.

Unfortunately, AC cannot be used to power solid-state devices such as transistors, integrated circuits, and diodes directly. So a power supply must not only be capable of providing the correct voltage and current, it must also convert the power into direct current (DC).

A typical power supply is shown in Figure 5-1. House voltage enters at point A at 120 V AC as represented by the AC waveform. The *step-down* transformer reduces the voltage at point B to what is required for that particular project. It is still AC, with a different voltage and/or current but the same frequency: 60 Hz. The rectifiers at point C convert the AC to pulsating direct current (PDC) at about the same voltage. Their job is to "fill in" the holes in the waveform. The PDC is smoothed out by a filter capacitor, C1, and a filter resistor, R1, at point D.

Of course, this is the raw base for a power supply. Other components are added to guarantee a constant source of current with regulation as well as a switch, a signal indicator, and a fuse, just a few of the additional parts that will make the supply more dependable and more sophisticated.

You can build any of the following power supplies that you need. Several are discussed so that you can choose

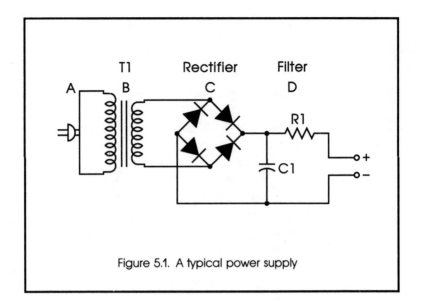

Figure 5.1. A typical power supply

which would be best for your purpose. I would definitely recommend building the 9-V battery eliminator if you have never built a power supply. It will teach you the principles of a power supply. No matter how complicated such a unit becomes, the basic elements are always the same.

9-VOLT BATTERY ELIMINATOR

Figure 5-2 shows a 9-V no-frills battery eliminator. You can use this power supply to operate any of the experiments that require no more than 9 V. Look at the other power supplies in this chapter, and as you study them you will see the resemblances among them.

Let's analyze what each component in this power supply does. From that experience you can assemble fancier and more complicated power supplies.

The transformer *steps* down the 120 V to 6.3 V.

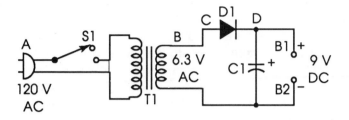

Parts List

B1, B2: Binding posts to match device to be powered

C1: 1,000 µF, 25 V electrolytic capacitor

D1: 50 V, 1 A rectifier diode

S1: Single pole, single throw switch

T1: 6.3 V AC, 0.5 A transformer

Note: The capacitor can be of a higher voltage as can the diode.

Figure 5.2. Schematic of no-frills 9-V battery eliminator

Safety Note:

Never plug the transformer into a wall outlet until certain precautions have been taken.

1. Connect the secondary of the transformer—these are usually green wires—to the input as marked on the circuit diagram. Connection is made as follows: Twist one green wire and one of the input circuit wires together tightly. Solder the pair. Now wrap electrical tape around the joint so that no bare wire is seen. Do the same with the second green wire and the other input circuit wire.

2. The black wires—the primary side of the transformer—must now be connected to a plug that will go into the wall outlet. These wires must be connected in the same fashion as you connected the green wires. Then join them to an electrical plug which will fit into the wall outlet.

3. Important: examine the connections to make certain they were made correctly. A minute or two now can save an entire circuit from burning, and will prevent you from being hurt. When you are *certain* you made the two connections correctly then and only then plug the transformer into the wall outlet.

4. After a few minutes of operation it is natural for a transformer to become warm to the touch. However if it becomes so hot you cannot keep your finger on the transformer casing—not the wires—you have something wrong. Unplug the transformer at once. Find your mistake before you go any further.

5. Never leave a transformer plugged into an outlet when someone is not there. It should be removed when you are through experimenting.

Without going into details, the diode converts the AC by a factor of 1.414 as it rectifies the current ($6.3 \times 1.414 = 8.9082$). We therefore obtain 8.9 V, which is close enough for our purpose. Capacitor C1 has the job of smoothing out the ripples in the current. P1 and P2 are binding posts or any type of connector that will fit the unit you plan to operate. Unless you know exactly what type of connector you will need, I suggest that you use four-way binding posts: one red and one black. *The red post will go to the positive side of the supply and the black to the negative or ground side.* Always use them that way. There is less chance of connecting a device incorrectly. These allow any type of connector to be attached.

First, we will assemble the parts on a breadboard. Look at Figure 5-2 to see how the parts are connected. The transformer is not set on the breadboard. Instead plug in wires from the secondary—the 9-V side—to the breadboard. **(Caution: Do not plug the transformer into the wall outlet until everything else is connected.)** The two wires to the primary side of the transformer should be soldered to insulated wires with a plug that will fit into the wall outlet.

Mount the switch on the breadboard and connect it between the wall outlet and the primary of the transformer. The positive side of capacitor C1 (marked with a plus (+) sign) is connected to the negative side of the diode and that in turn is connected to the positive output. This is one case where no resistor is employed. Remember that this circuit is very basic.

The previously unconnected side of the secondary is now attached to the negative side of C1, and that is connected to the negative output.

Before you do anything else, check your wiring carefully. Trace each connection from beginning to end. If you are certain you have made no mistakes, plug the transformer into the wall outlet and flip on S1. Don't leave it on long unless you have something that requires 9 V (such as a 9-V radio or cassette recorder) connected to the binding posts. Connect your voltmeter to the output. Use the range on your meter that covers *at least* 9 V. My meter has one range that goes up to 10 V. If your meter is like mine, that is the range you will use. The next range on my meter is 50 volts, but that will move the needle only slightly. *However, to prevent burning out your meter, always set it for a higher range than you think you will need.* Work your way down the ranges to protect your meter. When you get to a range that will move the needle enough so you are able to read the voltage, that is the range to use. If you have done everything correctly, your voltmeter should show about 9 V at the output.

If you want to make this project permanent, follow the wiring as you did on the breadboard and solder the connections. A piece of perforated board about 3 × 4 inches (7.5 × 10 cm) should do. Fit the parts into a plastic box about 4 × 5 (10 × 12.5 cm) inches with a hole for the wires to the transformer and two holes for the binding posts. To prevent the possibility of wires becoming frayed by rubbing on the edges of the hole, use a rubber gasket that fits around the hole. These are also sold in electronic stores. You can also mount the transformer on top of the box or choose a chassis large enough to fit the transformer inside. This transformer is quite small and requires little space.

The battery eliminator is fine for small projects that require no more than 9 V and less than 1/2 A. But other projects need power supplies that produce more voltage and are able to provide more power.

The answer is the next project. Although it can provide as much as 1 A and furnish any voltage from 5 to 20 V, it resembles the previous circuit. This power supply has one further advantage: it is *regulated*. What does this mean? It means that voltage remains constant no matter how much current is drawn until you reach the limit of one ampere. This is not the case for the 9-V battery eliminator. There the voltage fluctuates up and down, depending on how much power is required.

5-VOLT POWER SUPPLY[1]

The supply illustrated in Figure 5-3 is especially designed to operate a variety of integrated circuits (ICs) that require no more than 5 V. It is perfect when you need only 5 V and no more than 1 A. Look at the circuit starting from the left-hand side. One difference from the battery eliminator is that a fuse has been added between the 120-V input and the transformer. It's a good idea to use a fuse for any power supply. It does help to protect your unit in case of a project requiring too much current, which could burn out the diode. D1, D2, D3, and D4 make up a bridge rectifier which provides better rectification (changing AC to pulsating DC) than the single diode in the previous power supply. You can wire four diodes or buy one unit that incorporates all four in one case. There are more capacitors used than in the previous unit to smooth out the current. The power transistor, Q1, regulates the current so that it remains constant. Since Q1 will become quite warm as the current passes through it, a heat sink should be provided. Q1 is mounted on the metal chassis, but an *insulating kit* must be added. That consists of a piece of plastic with a heat-conducting compound. Every electronics store has such insulating kits. Note that the case of the transistor is the collector.

Z1 is a zener diode, which limits the current. (See the discussion of the zener diode in chapter 2.)

0- TO 15-VOLT REGULATED POWER SUPPLY

The unit shown in Figure 5-4 is different from the two previous projects in that you can vary the voltage from 0 V up to 15 V. The voltage is controlled by R2, a linear potentiometer (also called a "pot"). Current maximum is 1/2 A.

[1]Courtesy of GC Electronics.

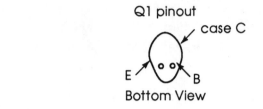

Q1 pinout

case C

E   B

Bottom View

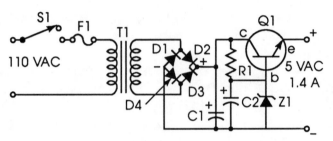

S1   F1   T1

110 VAC

D1   D2

D3

D4

R1

C1   C2   Z1

Q1

c   e

b

5 VAC

1.4 A

Parts List

Calectro
Cat No.

S1: SPST toggle switch   E2-130
C1: 2,200 µF, 25 V electrolytic capacitor   A1-134
C2: 47 µF, 25 V electrolytic capacitor   A1-129
C3: 4.7 µF, 25 V electrolytic capacitor   A1-126
D1, D2, D3, D4: 1 A, 50 V silicon diode   J4-1640
F1: 3A2GA quick acting fuse 2 amps.   D2-132
T1: 12.6 V AC, 2 A transformer   D1-747
R1: 150 Ω, 2 W resistor   26-428
Q1: NPN silicon power transistor   J4-1640
Z1: 5.1 V, 1 W zener diode   JR-1615

Fuse holder, heat sink for Q1, perforated
board, etc.

Figure 5.3. Schematic of 5-volt power supply

A typical power supply.
Note the on-off switch and fuse
to the side, set out of the way.
Connectors, meters, and controls
are on the front panel for easy use.

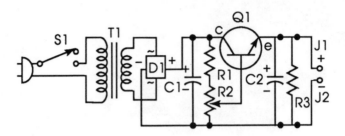

Parts List

C1: 2,500 µF, 35 V capacitor
C2: 250 µF, 35 V capacitor
D1: Bridge rectifier
     50 V @ 0.5 A
Q1: 2N3055 transistor
R1, R3: 1 K, 1 W carbon resistor
R2: 10 K linear potentiometer
S1: Slide switch
T1 120 to 24 V transformer, 1.0 A

Figure 5.4. Schematic of 0-15 volt
regulated power supply

Again you can substitute a solid-state bridge rectifier for
four diodes. Q1, a power transistor, acts as a switching device. As you can see, this supply is not very different from
the previous unit.

　After you mount the variable potentiometer, glue a
piece of cardboard to the chassis and, using a meter, make
a scale showing the various voltage levels. Figure 5-5 shows
a typical calibrated dial.

　The way to calibrate the dial is to use the VOM across
the outputs. Cut out the smaller circle and fit it around the

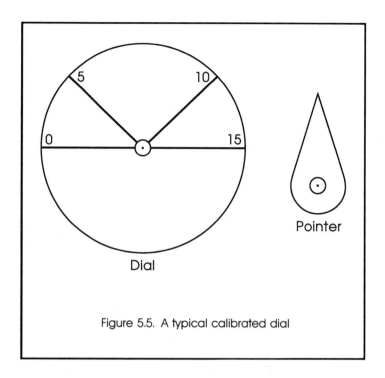

Figure 5.5. A typical calibrated dial

shaft of the pot. Glue it in place. The pointer is glued to the knob that fits on the shaft.

Set the potentiometer all the way down so that your meter reads 0 V. Make a mark on the dial and label it 0. Crank up your voltage until your meter reads 15 V. It's best to provide a *load* (e.g., a 10K resistor at 2 watts). Pencil that mark on your dial. If your potentiometer is perfectly linear, you divide the space from 0 to 15 in equal parts. You might want to mark off every volt or every other volt or any other way you wish. Place the knob so that the pointer is at the zero (0) point and tighten it in place.

The same idea will work for any power supply.

Instead of the calibrated pot, you can connect a 0- to 25-V meter across the output so that you will know exactly what voltage you are producing at any given time. Meters are available in all electronic supply houses.

6

EXPERIMENTS WITH A
TIMER-INTEGRATED CIRCUIT

THE 555 INTEGRATED CIRCUIT

The integrated circuit (IC) you're going to be experimenting with in this chapter is so popular and so useful—so
many things can be done with it—that whole books have
been written about it. The reason for its popularity is that
apart from being able to do so many things, it is not delicate and allows for a great deal of experimenting with
little danger of burning it out. Naturally you have to stay
within its voltage limits. You must provide no less than 4.5
V and no more than 16 V—DC, of course. A good safe
operating voltage range is 5 to 9 V.

The IC in question is the 555, along with its big brother,
the dual version, the 556. Figure 6-1 shows the "pinouts"
for the little rectangular 8-pin type known as the *V package*, as well as for the 14-pin dual 555 called the 556.

It is a timing circuit, is also able to produce square
waves, and has lots of other uses. Many companies man-
ufacture the 555. They each use their own parts number,
but all are basically the same.

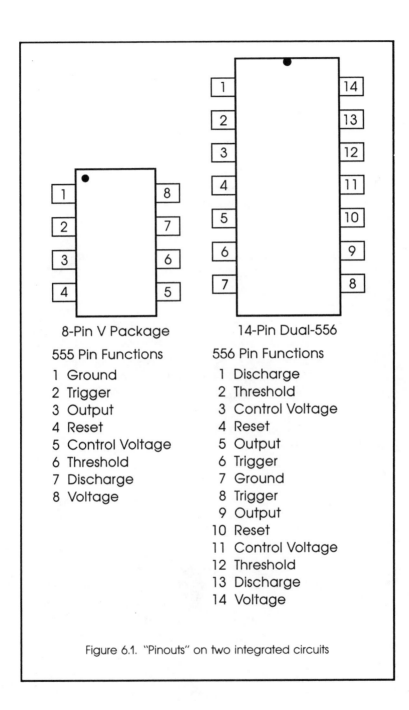

8-Pin V Package

555 Pin Functions

1 Ground
2 Trigger
3 Output
4 Reset
5 Control Voltage
6 Threshold
7 Discharge
8 Voltage

14-Pin Dual-556

556 Pin Functions

1 Discharge
2 Threshold
3 Control Voltage
4 Reset
5 Output
6 Trigger
7 Ground
8 Trigger
9 Output
10 Reset
11 Control Voltage
12 Threshold
13 Discharge
14 Voltage

Figure 6.1. "Pinouts" on two integrated circuits

Inside the single version are packaged some twenty transistors, fifteen resistors, and two diodes. Imagine if you had to wire all those components together: there would be almost one hundred solder joints! Instead the 555 has only 8 pins to connect and 14 pins for the dual version. The dual version contains two complete 555s. Both versions are illustrated in Figure 6-1.

No wonder it's been so popular. You can experiment to your heart's delight. When you breadboard it, use an 8-pin or a 16-pin socket. Do the same thing when you hard-wire it. In fact, it's a good idea to do that with every integrated circuit you use. You don't have to bend the delicate leads of the IC, and soldering them can be a very tricky proposition. Hold the soldering gun too long on a lead and you find yourself with a dead unit. Use a *vector board* when you use ICs with or without sockets. This type of board has the holes placed close together so that you slip the socket right into the holes. Wire wrapping is the best way to go when you have a couple of ICs on a board. Then, when your circuit is wired and you are certain that everything is where it should be, you can plug in the IC—but not before.

The 555 or 556 is perfect for any portable application, as you can run it with three 1.5-V batteries, such as AAs or even AAAs. I've used it as a portable signaling device with only one 9-V battery. Even when the batteries begin to run down, the 555 runs just as well as ever. Of course, you can use any of the power supplies you've already built, so long as you stay within the voltage limits.

It operates in two modes, depending on how it is wired. In the *monostable mode* it's a one-shot multivibrator. As an *astable multivibrator* it's a square-wave clock. A *monostable multivibrator* is a circuit that has only one stable state. It can be triggered, however, to change its state, but only for an interval. This interval is set by a couple of external components. After that interval ends, the unit returns to its original state.

As an *astable multivibrator,* the 555 operates as a circuit that has two momentary stable states. It alternates rapidly from one state to the other for a period which is also set by external components.

In the next project you will operate the 555 as a one-shot multivibrator.

AUDIBLE TESTER

Let's say that you are testing the continuity of a circuit but you cannot see the light-emitting diode (LED) that indicates whether things are right. And if you're using your VOM, it turns out you can't see the needle. Here's the solution: Build the audible tester.

The best way is to wire it on a piece of perforated board with a socket for the IC and the smallest capacitors and resistors you can get. (See Figure 6-2.) Buy 1/4-W carbons, if you can. Otherwise the 1/2-W will do. Mount the components vertically; that way you can package the whole thing in a small box that you strap to your belt. I used a plastic chassis measuring 4½ × 2½ × 1½ inches (11 × 6 × 4 cm).

You have probably noticed that there is no switch. It's not necessary since the unit draws no power until it is testing a good connection. I used a 9-V battery and attached it by means of a snap connector. The Sonalert is one of the many piezobuzzers now available. They give off a good loud sound and draw little current. You can also try a tiny speaker.

Use four-way binding posts. That way you can connect any prods you own. Terminate the prods with alligator clips or any clips that will hold a wire firmly while you are testing.

Drill holes on the cover for the sounder and for the two binding posts. Polarity is not important, as *no current must be flowing through the circuit while you are testing.*

Once you've wired up the unit test for continuity by

**A continuity tester: All of the
parts can be packed
in a tiny plastic box.**

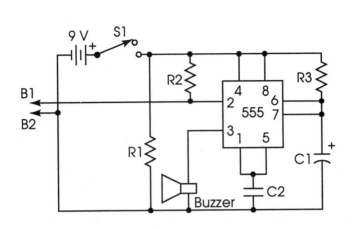

Parts List

B1, B2: Binding posts
Buzzer: Any 5 V piezo-type buzzer or speaker
C1: 0.1 µF, 10 V electrolytic capacitor
C2: 0.01 µF disk capacitor
R1: 1 KΩ, $\frac{1}{2}$ W resistor
R2, R3: 100 KΩ, $\frac{1}{2}$ W resistor
S1: SPST Switch

Figure 6.2. Schematic of the audible tester

touching the two prods together, you should hear a sound if there is continuity.

GENERAL-PURPOSE ALARM

To convert the IC into an alarm, we are going to make it operate once again in the monostable state. In other words, it will be quiet until it is triggered to sound off. But the beauty of the circuit is that by shifting the position of switch 1 we get exactly the reverse condition. Moving the switch ensures that the alarm is on except when the circuit is closed (see Figure 6-3).

Notice the few parts that have to be added to the 555. Second, look at points A and B. If switch S2 is in the top position and the two prods touch each other, a loud shriek comes out of the buzzer. That's similar to the way we used it as a tester in our last experiment. But it does more than that: it can also be used as a bilge pump alarm or as a warning alarm whenever a liquid rises up to a certain level.

How does it work? Very simply. Water and most liquids conduct electricity. That means that if both prods touch water, it will have the same effect as if the two prods touched each other. And when they do, the alarm sounds.

The unit itself should be enclosed within a chassis; plastic is fine. Connect the two prods at the end of two long insulated test lines. Tie the two prods together, separating them with a piece of plastic so they are about a quarter-inch away from each other. Make them secure. Now suspend the prods so that they are just above the warning line. When the water rises at the bottom of a boat or in a tank and reaches the prods, the alarm goes off. No spark is generated when the alarm goes off, so there is no danger that a spark will ignite gasoline or another volatile explosive liquid.

The unit can operate with a 9-V battery. But since you need current to be applied continuously it would be best to use the 9-V battery eliminator (see Figure 5-2).

This alarm is based on the idea that switch S2 is in the upper position. However, if the switch is moved to the lower

(Facing page) Figure 6.3. Schematic of the
general-purpose alarm

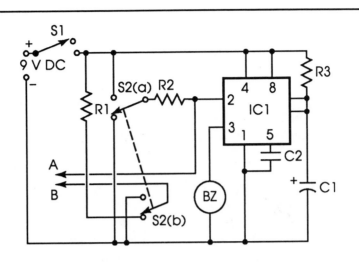

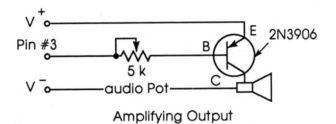

Amplifying Output

Parts List

BZ: Piezobuzzer
C1: 2 µF, 15 V capacitor
C2: 0.01 µF disk capacitor
IC1: 555 or similar
R1: 1 KΩ, $\frac{1}{2}$ W resistor
R2: 10 KΩ, $\frac{1}{2}$ W resistor
R3: 170 KΩ, $\frac{1}{2}$ W resistor
S1: SPST switch
S2: DPDT switch
Plastic case, prod, connectors,
test cords, IC socket

position, the alarm goes off only when the circuit is broken. Now you can adapt the circuit to shriek when a window or door is opened. Point A, for example, could be fastened to the window and next to it on the window frame would be attached point B. A piece of copper large enough to touch both prods would be fastened to one or the other. As long as the copper is touching both prods, the alarm is silent. But if a thief unknowingly breaks the connection——!

Switch S1 disconnects the alarm regardless of the setting of S2. This allows you to open a window or a door without alerting anyone.

In case you must position the alarm where you are afraid it cannot be heard, you might add the circuit labeled "AMPLIFYING OUTPUT." This is a one-transistor amplifier using a 2n3906 transistor. You substitute this for the buzzer and add a small speaker. Several stores sell a small amplifier. Some even have a built-in speaker. In that case, you can ignore the following explanation of the circuitry for the amplifier.

The emitter of the transistor is connected to the positive side of the power supply. A wire from pin no. 3 goes to the 5K pot: the volume control. The other side of the resistor is connected to the base of the transistor. The collector goes to the speaker, and the other side of the speaker is connected to the negative side of the power supply. Run the speaker from the amplifier by means of ordinary line wires.

TWO-TONE WARBLER

Here's how to use a 555 in still another way: as a two-tone warbler. For this we use two 555s; they are illustrated in Figure 6-4A. You can substitute one 556 for the two ICs (illustrated in Figure 6-4B) to make the device more compact.

When you see an IC marked ½IC1 in two different

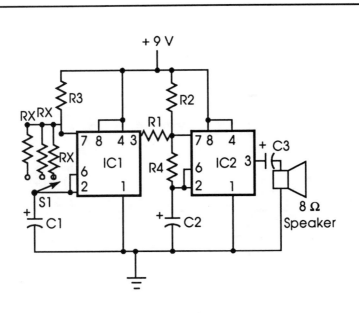

Parts List

C1, C2: 0.1 µF to 1.0 µF
C3: 10 µF
R1: 100 k
R2: 250 k to 330 k
R3: 10 k
R4: 10 k to 15 k
RX 550 k to 1 MΩ
S1: Rotary switch (see text)
All resistors from $\frac{1}{2}$ to $\frac{1}{4}$ W

Figure 6.4A. Schematic for two-tone warbler
using two 555 integrated circuits

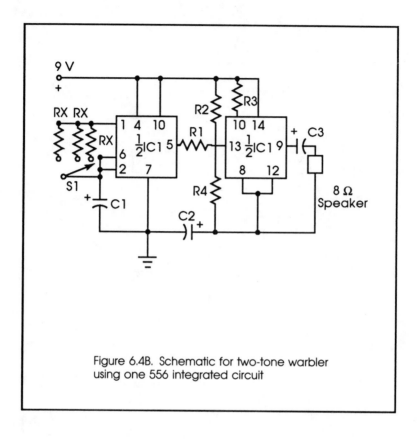

Figure 6.4B. Schematic for two-tone warbler using one 556 integrated circuit

places in a schematic, that doesn't mean that the two halves are physically separated. They are drawn that way to make it simpler to understand.

For example, in Figure 6-1 you notice that there is one common ground for both halves of the IC. There is also one pin (pin 14) for the voltage input. However, there are two control voltage points (pins 3 and 11), just as there are two outputs (pins 5 and 9). Output of the first half of the IC is pin 5. This is connected to the second half—to pin 13 via R1. The output of the second half is pin 9, which feeds into the speaker. The other pins are connected in the same way.

You will encounter not only dual versions of some ICs but even some *quadruple versions:* four parts of the same IC. Each section operates as a completely separate unit. You can see how this simplifies the building of a complicated circuit. At the same time the double or quadruple IC requires less room.

The basic difference in this project is that you are making one 555 (or one-half of the 556) produce one tone and the second, or the other half, produce a second tone—all this from one power supply. This is a circuit that allows for much experimenting. To vary the tone of IC1, change the values of C1, C2, and Rx. Rx can be different resistors mounted on a switch. You can also use a linear pot with a value of about 1 megaohm (MΩ). As you rotate the switch, you are introducing different values for the resistance, and this enables you to vary the tone. The best way to produce the tones for your purpose is to breadboard the entire unit. This permits as much experimentation as you wish.

You can vary C1 and C2 from 0.1 to 1.0 μF. The group of Rx resistors can go all the way up to 1 MΩ. You can also experiment with R2 and R4. You must keep the two tones far apart in frequency; otherwise you lose the effect of two separate tones.

TOUCH SWITCH

To turn an appliance on or off you usually have to press or turn a switch in some mechanical way. This project is an unusual switch. All you have to do is touch it with your finger and the light or whatever else you want to control goes on.

How does it work? The circuit is based on our old standby, the 555. As you can see in Figure 6-5, the IC is wired as a monostable. The difference is in the method of triggering the device. The trigger input, pin 2, is biased to a high value by the 22Ω resistor, R2. Touch the contact

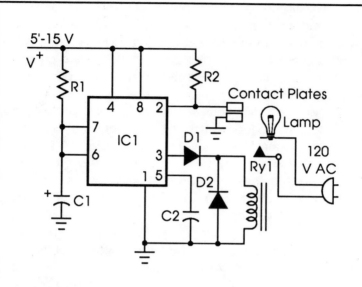

Parts List
C1: 1 μF, 25 V DC
C2: 0.1 μF
D1, D2: 1N914
R1: 4.7 M Ω
R2: 22 M Ω
Ry1: Relay 6 V, 500 Ω Approx. 12 mA

Figure 6.5. Schematic for the touch switch

plates and your body resistance lowers the *impedance* (the combination of resistance and capacitative react-ance that offers resistance to an alternating current) from pin 2 to ground. This reduces the voltage and the timer will start. For the components in the present schematic the output pulse width is about 5 seconds.

Connect a relay from pin 3 to ground. This will control whatever you wish. The output pulse is just sufficient to trigger the relay. What's great about a relay is that a tiny voltage triggers it, and that in turn controls something that requires a large voltage. This way the larger voltage does not come into contact in any way with the 555, which is operating at a few volts.

The contact plates should be made of copper or brass about 1/2-inch long and about the same width. Fasten them on a base so that they are about 1/16-inch apart. This way your one finger touches both plates at once.

7

MORE ADVANCED EXPERIMENTS WITH INTEGRATED CIRCUITS

INTERCOM[1]

The intercom in Figure 7-1 is a project you will enjoy building and using. It can be connected between the house and the basement workshop or the garage. You can use it as an electronic baby-sitter. The heart of the unit is an integrated circuit, the LM380. This IC is a low-noise dual amplifier. Internally it contains twelve transistors, seven resistors, and a few other parts. That simplifies much of your work. All you have to do is add a few components to produce a very useful project.

It operates on a wide range of voltages from 10 to 22 V DC. I have chosen 15 V as it lies well within the working range and you have probably built a power supply that will furnish the required voltage and current.

The unit works in the *half-duplex mode*. That means that while one person is talking, the other must listen. It is not like a telephone, which operates in the full-duplex

[1]Circuit courtesy of National Semiconductor.

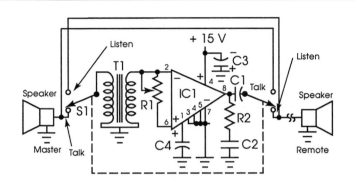

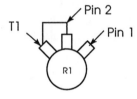

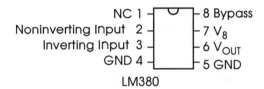

NC 1 ― 8 Bypass
Noninverting Input 2 ― 7 V₈
Inverting Input 3 ― 6 V_OUT
GND 4 ― 5 GND

LM380

Parts List

C1: 50 μF capacitor at 25 V
C2: 0.1 μF disk capacitor
C3: 200 μF capacitor
C4:, C5: 5 μF capacitor
IC1 Lm380 integrated circuit
R1: 2.5 m Ω audio potentiometer
R2: 2.7 Ω, $\frac{1}{2}$ W resistor
S1: DPDT switch
Spkr 1, 2: small 8 Ω speakers
T1: 8 Ω/200 Ω audio transformer

Figure 7.1. Schematic for the intercom

mode. The telephone allows you to talk at the same time the other person is talking. Depending on the position of switch 1, you either talk or listen. Once the switch is moved, the listener becomes the talker, and vice versa.

You are probably wondering where the microphones are located. There aren't any. The small speakers—they should be no more than 2 or 3 inches (5 or 7 cm) in diameter—act as microphones, depending on the position of the switch.

The switch, a double-pole, double-throw switch (2P2T), is located at the master station, as are most of the other parts. The only exceptions are the remote speaker and the line connecting it. You can place the remote as far as 50 feet (15 m) from the master station. Use speaker wire or even ordinary line cord for the connections between the master and the remote.

One side of the cord acts as the ground, and all ground connections should be made to one common point within the master station. This can be housed in a plastic or metal box with an opening for the speaker and another for the switch handle.

R1 is the volume control. The resistance is varied depending on the distance between the two stations and the quality of the speakers. The unit is not designed to be used outdoors unless the wires are heavily insulated and the remote station is sheltered from the elements.

Once the volume level has been tested, you will not need to vary it again.

A NAND GATE SIREN [2]

What is a NAND gate? First let's define what a gate is. An electronic gate is the equivalent of an ordinary gate. It's a door; instead of using a key to open or close an electronic gate, you use an electronic code. The code must

[2] Circuit courtesy of GC Electronics.

be exact to control the gate. Four different types of gates are illustrated with their respective codes and symbols in Figure 7-2. They are *OR gates, AND gates, NOR gates,* and *NAND gates.*

Gates may have two, three, four, or eight inputs. Regardless of the number of inputs, the code is always the same. Illustrated in Figure 7-3 are the schematic and the pinout for the J4-1000. (This is GC Electronics's own part number; the generic number is 7400.) This particular IC is a quadruple two-input positive NAND gate.

Pin 14 is the common voltage input for each IC. Pin 7 is the common ground. Pins 1, 2, 4, 5, 9, 10, 12, and 13 are

SYMBOL	CODE
OUTPUT INPUTS "AND" GATE"	Output is "on" only when all inputs are "on" Output is "off" when any or all inputs are "off".
OUTPUT INPUTS "NAND" GATE"	Output is "off" only when all inputs are "on" Output is "on" when any or all inputs are "off".
OUTPUT INPUTS "NOR" GATE"	Output is "on" only when all inputs are "off" Output is "off" when any or all inputs are "on".
OUTPUT INPUTS "OR" GATE"	Output is "off" only when all inputs are "off" Output is "on" when any or all of the inputs are "on".

Figure 7.2. Symbols for AND, NAND, NOR, and OR gates

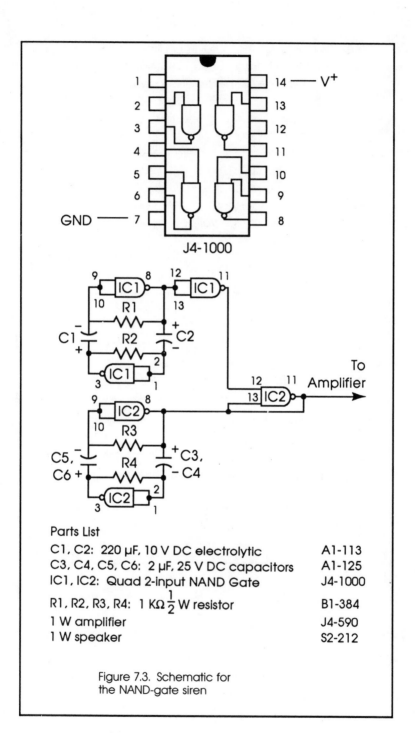

Parts List

C1, C2: 220 µF, 10 V DC electrolytic	A1-113
C3, C4, C5, C6: 2 µF, 25 V DC capacitors	A1-125
IC1, IC2: Quad 2-input NAND Gate	J4-1000
R1, R2, R3, R4: 1 KΩ $\frac{1}{2}$ W resistor	B1-384
1 W amplifier	J4-590
1 W speaker	S2-212

Figure 7.3. Schematic for
the NAND-gate siren

the inputs, and pins 3, 6, 8, and 11 are the outputs. Each IC uses only three of its outputs and six of its inputs.

To turn an input on, connect it to 5 V DC through a 1-kΩ resistor and a switch. Another way to turn the input on is to connect it directly to the output of another gate or an integrated circuit. To turn the input off, you connect it to ground directly or via a switch, or to the output of another gate or integrated circuit.

Remember that any gate that is not connected to ground or to a 5-V source becomes on. Whether the input is used or not, it must be connected either to ground or to the voltage source.

So how does this gate operate? Its output is off only when all inputs are on. The output is on when any or all of the inputs are off.

The siren project depends on two quadruple two-input NAND gate ICs. They operate as multivibrators. One operates at a slow rate and the other at an audio rate (much faster).

The faster multivibrator (IC2) is switched on and off by the slower multivibrator (IC1).

You can apply any voltage between 4.75 and 5.25 V DC. The voltage must not exceed 5.5 V. To be perfectly safe, it is best that the power supply should not produce more than 5.0 V. You can use any of the variable power supplies you have built. The complete siren system is shown in Figure 7-4. The system merits some study. It helps you to understand the relations among the power supply, the siren module, and the amplifier.

In Figure 7-3 look at C3, C4, C5, and C6. They are connected in series. This means that the negative side of C3 is connected to the positive end of C4. Since each

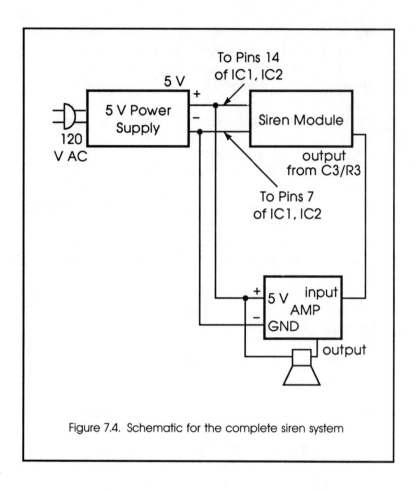

Figure 7.4. Schematic for the complete siren system

capacitor is rated at 2 μF and they are in series, what is the total value of each pair of capacitors? Use the formula in Figure 2-1C. Try the equation yourself. The answer should be 1 μF. If you can find such a unit, get one for each pair of series capacitors. Do the same with C5 and C6.

Use the perforated vector board with its holes that match the IC socket pins. If you prefer to solder the IC, it must be done very, very carefully. But it's much safer to use sockets for the ICs. You may want to change an IC for some reason, and it is difficult, if not almost impossible, to unsolder an IC without burning it out, even with a great deal of care.

For any circuit that has more than one IC, it is best to

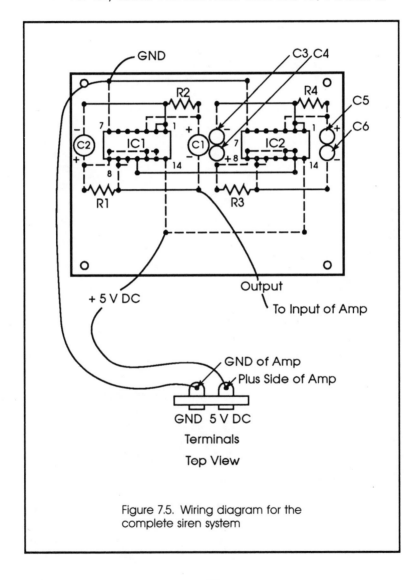

Figure 7.5. Wiring diagram for the complete siren system

use thin wire—no. 26 buss—to make the connections. Wire wrapping is the easiest way to wire the circuit when you have two ICs, as in this project. Buy a little wire-wrapping tool and a spool of the special wire that is made for wire wrapping. It ensures a neater job and prevents soldering in tight places. Again, don't insert the ICs into the sockets until you have finished your wiring and checked for any mistakes. Then attach the power supply to the board. The wiring diagram is shown in Figure 7-5 to make this circuit easier to wire.

8

HIGH-TECH DICE AND LIGHT DIMMER

LIGHT DIMMER

Safety Note:
Since this circuit handles house voltage, this should be done under adult supervision.

A light dimmer can be an aid to your photography if you use a couple of large spotlights or 300-W floodlights when shooting your favorite subject. If you leave the lights on full when you are focusing or simply adjusting the lights, the room becomes warm, your model begins to sweat, and you shorten the life of the lights. The solution is the simple dimmer described next. Look at Figure 8-1; you can see that few parts are necessary.

There are other uses for this handy dimmer. I mounted mine on a shaft of a powerful desk lamp. I find that most of the time I use the light at about only one-third of its full

This dimmer switch can be attached
to a lamp or any device operating within
the rated power of the triac.

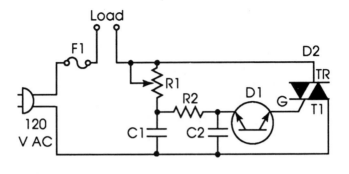

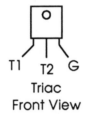

Triac
Front View

Parts List
C1: 0.1 µF, 400 V
C2: 0.01 µF, 400 V
D1: Trigger Diode HEP 311
D2: Triac HEP R1243
F1: 4A Slo-blo fuse
R1: 200 K 2 W Linear Pot
R2: 4.7 K 1 W
AC jack, chassis, gasket for line cord entry

Figure 8.1. Schematic for the light dimmer

output. I turn it on full only when I am doing delicate work such as wiring tiny components. As a result the lamp has lasted several years. The dimmer gives you between 5 and 95 percent of a lamp's full output.

As you can see in Figure 8-1, the potentiometer (R1) allows you to adjust the light to suit your needs exactly. D1 in the schematic looks like a transistor, but in practice it behaves like two diodes back to back. The triac, D2, is the engineer; it regulates the amount of light on the basis of the orders it gets from the potentiometer. A total of 800 W can be handled with this dimmer. That means a 500-W spotlight and up to three 100-W lights all on the same line. It's better to stay under the limit of the wattage to prevent the unit from overheating.

Everything fits into a metal chassis $4 \times 2 \times 1\frac{1}{2}$ inches ($10 \times 5 \times 3.75$ cm). Drill a hole large enough so that you can fit a light cord through it. But first place a rubber gasket so that the cord does not become frayed from the sharp edges of the hole. Fasten the light cord inside with a knot so that no pulling will rip out the cord. Use heavy insulated wire to connect the parts. A piece of perforated board about 3 inches long and 3/4 inch (7.5×1.5 cm) wide fits very nicely into the box. Use rubber feet under the board so that none of the connections touches the chassis. You can also "hang" the components from insulated connector points instead of the perforated board. Just make certain that no bare wire touches the chassis.

Drill a hole into which to fit an outlet for whatever you want to connect. The shaft of the potentiometer fits into the last hole you need to drill. Add a knob to the shaft, and that's all there is to it. There is no need for calibration since you adjust the lamp to suit your needs. The metal chassis can be covered with contact paper to make it look like grain, or you can paint it any color you want so it will look like a store-bought unit. There is no reason for homemade projects to look unfinished and less attractive than those you buy.

A PAIR OF ELECTRONIC DICE[1]

Safety Note:
No more than 5 V should be used with this project.

Many board games such as Monopoly and Parcheesi require a pair of dice. Here's how you can bring those old favorites right into the high-tech world. In the last chapter you gained some experience with a logic gate. We are going to use the same one: a quadruple two-input positive NAND gate. It will create the same haphazard patterns as a pair of the ivory dice. That means that you will never know what combination of numbers is going to come up. The complete schematic is shown in Figure 8-2.

To imitate the spots on the dice, fourteen light-emitting diodes (LEDs) are placed in the arrangement shown in Figure 8-3A. But first let's examine how the circuit works. Half of one NAND gate operates as a clock and forms a wave-shaping circuit. (See the discussion of NAND in Chapter 7.) This circuit drives a divide-by-6 counter consisting of one and one-half of a dual J-K master-slave flip-flop. These flip-flops drive a second divide-by-six counter.

In this circuit the two flip-flops act as counters and as dividers. They count in a way that is peculiar to them. A flip-flop counts 1, 2, 4, 8. By adding these four numbers in different combinations, you can arrive at every number up to 15. However, for our purposes we will go only as far as 12. Here's an example: To count to 3, outputs numbered 1 and 2 will be "on." A 5 is a combination of 4 and 1. The outputs of the counter are connected to an IC, which decodes the numbers to our own form of counting: 1, 2, 3, 4, and so on.

The decoding is done by another quadruple two-input positive NAND gate with the help of a hex inverter. This

[1]Courtesy of GC Electronics.

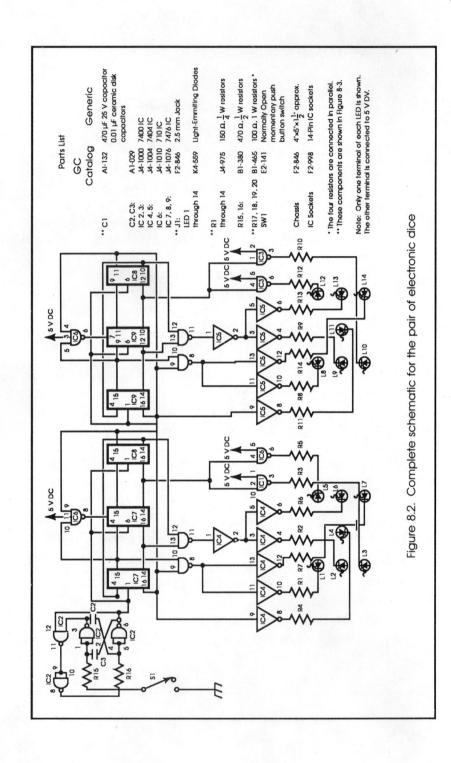

Figure 8.2. Complete schematic for the pair of electronic dice

Parts List

	GC Catalog	Generic
** C1	A1-132	470 μF 25 V capacitor
	A1-029	0.01 μF ceramic disk capacitors
C2, C3:	J4-1000	7400 IC
IC 2, 3:	J4-1004	7404 IC
IC 4, 5:	J4-1010	7410 IC
IC 6:	J4-1076	7476 IC
IC 7, 8, 9:	F2-846	2.5 mm Jack
** J1:		
LED 1 through 14	K4-559	Light-Emmiting Diodes
** R1 through 14	J4-975	150 Ω, ¼ W resistors
R15, 16:	B1-380	470 Ω, ½ W resistors
** R17, 18, 19, 20	B1-465	100 Ω, 1 W resistors *
SW1	E2-141	Normally Open momentary push button switch
Chassis	F2-846	4"×5"×1½" approx.
IC Sockets	F2-998	14-Pin IC sockets

* The four resistors are connected in parallel.
** These components are shown in Figure 8.3.

Note: Only one terminal of each LED is shown.
The other terminal is connected to 5 DV.

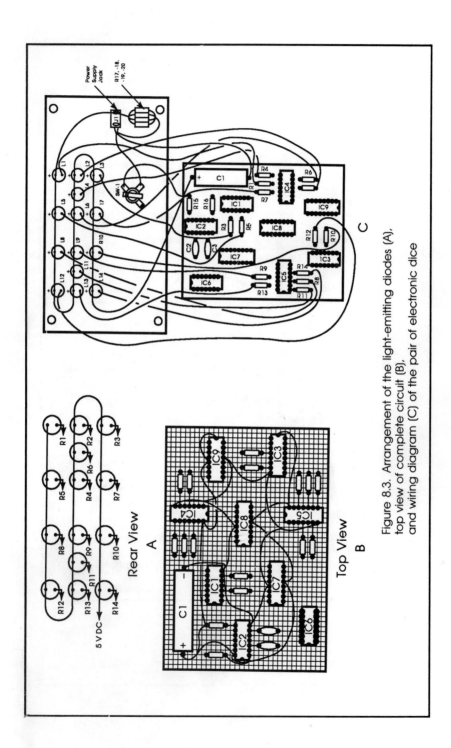

Figure 8.3. Arrangement of the light-emitting diodes (A), top view of complete circuit (B), and wiring diagram (C) of the pair of electronic dice

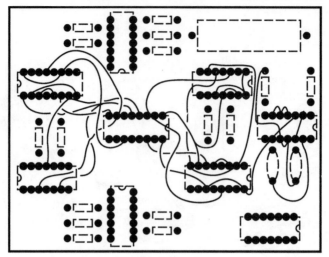

Bottom View

A

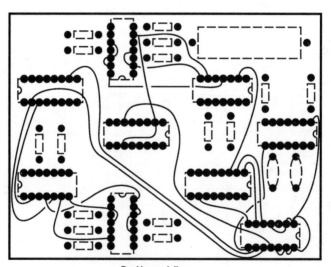

Bottom View

B

Figure 8.4. Two steps in wiring the electronic dice

drives the seven LEDs for each dice with a current-limiting resistor. The extra triple three-input positive NAND gate, IC6, serves to turn off the LEDs when power is first turned on. This prepares them to operate randomly.

Each LED has two leads. It doesn't matter which way you connect them. The limiting resistor, 150 Ω, will prevent the LED from burning out even if you connect one the wrong way. They just won't light if they are connected incorrectly.

CONSTRUCTION

Study the complete circuit shown in Figure 8-3. It is more complicated than those you've worked on before. Go over it carefully before you start wiring. Your understanding of what goes where will make it easier when you begin to wire the connections from component to component. To help you, the two bottom views show the details of the wiring from IC to IC. The top view is shown in Figure 8-3B.

The twelve LEDs are mounted on the face of the chassis or on the top if you prefer. The case may be metal or plastic. Separate the two groups from each other so that they look like the faces of two separate dice. By all means use a wire-wrapping tool to make the connections between ICs. And don't forget to use IC sockets mounted on vector board. To make your wiring easier follow the example shown on the bottom of Figure 8-3C.

Figure 8-4A and B shows two steps of the wiring for the bottom view. Start your wiring by following one or the other diagram first. It doesn't make any difference with which one you begin. Check one thoroughly before you go on to the next. Notice the notch at the top of each IC. Be sure to place the ICs the same way into the sockets.

You can use any of the power supplies you have built, provided *you do not exceed 5 V.*

Set up whatever board game you plan to play. Press the switch and say good-bye to dice that roll off the table.

9

HOW TO BUILD METERS FOR ADVANCED MEASUREMENTS

VERY LOW OHM METER

Resistors are so important in electronics that it is not surprising that there are several ways of measuring their resistance very accurately. The project that follows can measure even a fraction of an ohm.

When do you need such accuracy? Let's say that after you have soldered two or more wires together you find that the circuit is not operating quite the way it should. After testing various components you wonder whether that last solder joint could be at fault. A solder joint may throw off a delicate circuit because the solder is increasing the resistance. What's even more common is a cold solder joint. Your VOM can measure low ohms, but at the bottom of the scale it is not accurate enough to have any value.

The project (Figure 9-1) uses your VOM, but instead of measuring ohms directly you are taking advantage of the 50-μA scale. The circuit is a simple one that can be built inside a small chassis. S2 is a double-pole normally open switch. Q1 is any PNP transistor. Buy the 1K potentiometer

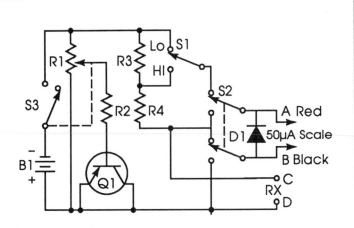

Parts List
B1: 3 V battery
D1: 20 V silicon rectifier
Q1: any pnp transistor
R1: 1 K Ω linear potentiometer
R2: 150 Ω carbon resistor
R3: 4 ohm 10% carbon resistor
R4: 1 ohm 10% carbon resistor
S1: SPDT switch
S2: Momentary open 2P2T switch
S3: SPST switch
4 Binding posts

Figure 9.1. Schematic of the very low ohm meter

with a switch, which then becomes S3. That saves you some space. Mount the three switches, the potentiometer, and four binding posts on the front panel. Insert the leads from the VOM into one red and one black binding post. The other two posts which have no polarity can be any color you choose. Label them at once so you do not mistake their function. You will be using the 50-μA scale of the meter. Using a piece of wire, temporarily connect binding posts C and D. Depress S2 while rotating R1 until you get a zero deflection on the 50-μA scale. Do this with both the low and high settings of S1. This is necessary because as the battery ages it will throw off your measurement unless you reset R1 just as you are about to use the meter. Start off by setting S1 at its high setting. Attach your unknown resistance to points C and D. Press S2. If you get no reading, then change the setting for S1. The reading you get on the VOM is in ohms.

SOLID-STATE
CAPACITANCE METER

Suppose, like most experimenters, you have picked up a quantity of surplus capacitors and because of the complicated codes that manufacturers mark on their units you don't know their value.

It's difficult to figure out what dot in the six- or nine-dot code means what. Eventually you do what many experimenters do—put them away. So they lie there doing nothing. The solution to these mysterious unknowns is the solid-state capacitance meter shown in Figure 9-2.

The meter will measure all capacitors, regardless of their type, from 100 pF to 10 μF, with a remarkable degree of accuracy. This range covers practically all capacitors you may have.

The two ICs are available almost everywhere, and the circuit is not a difficult one to assemble. Use 14-pin sockets for the ICs. The resistors, R3, R4, R5, R6, R7, and R8, are tiny potentiometers with an adjustable screw so you can vary their value. This is necessary when you first calibrate your meter.

You need six test capacitors. They must have a 2 percent tolerance, as they will affect the actual measure-

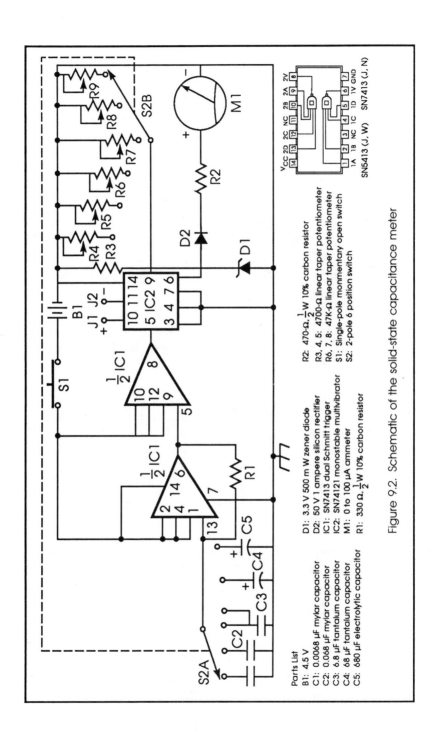

Figure 9.2. Schematic of the solid-state capacitance meter

Parts List

B1: 4.5 V
C1: 0.0068 μF mylar capacitor
C2: 0.068 μF mylar capacitor
C3: 6.8 μF tantalum capacitor
C4: 68 μF tantalum capacitor
C5: 680 μF electrolytic capacitor

D1: 3.3 V 500 m W zener diode
D2: 50 V 1 ampere silicon rectifier
IC1: SN7413 dual Schmitt trigger
IC2: SN74121 monostable multivibrator
M1: 0 to 100 μA ammeter
R1: 330 Ω, $\frac{1}{2}$ W 10% carbon resistor

R2: 470-Ω, $\frac{1}{2}$ W 10% carbon resistor
R3, 4, 5: 4700-Ω linear taper potentiometer
R6, 7, 8: 47K-Ω linear taper potentiometer
S1: Single-pole momentary open switch
S2: 2-pole 6 position switch

ments of your unknown capacitors. The more accurate the test units are, the more accurate will be your readings. It is also important that capacitors C1, C2, C3, C4, and C5 be close tolerance units with low leakage.

Switch S2 has two plates. To one you wire the resistors, and to the other, the capacitors. Make certain that when the switch is making contact with R1 it is also making contact with C1. Mount all the parts on the switch before you install it on the chassis. The size of the chassis depends on the size of the meter you get. On the front of the chassis should be the meter, the binding posts, and the range switch. Mount the momentary switch on top of the case. You can also mount it on the front panel, except that when you depress the switch the case will move. Label each position of the range switch.

Do not close the chassis yet. First is the calibration. J1 and J2 are binding posts to which you connect the test capacitors. Polarity is unimportant except with electrolytics, but as was mentioned before, you will usually not be testing those. When you do connect an electrolytic, make certain that the positive end is placed in J1, the red post.

Turn to the 10-μF test capacitor. Depress the momentary switch while you adjust R4. Turn the screw on the pot slowly until the meter needle is fully deflected. Take the next capacitor, 1 μF, and adjust R5. Do this for every range.

When you are testing an unknown capacitor, clip it to the binding posts and start with the highest range and work your way downward, depressing S1 on each range, until you get a deflection on the meter. Read the position of the meter needle: that is, the value of the capacitor. After finding the value of a capacitor, mark that value on a piece of tape and put it on the capacitor. Remove it only when you wire the capacitor in a circuit.

HIGH OHM METER

If you don't own a VOM and you need to measure a large resistance, then an ohmmeter becomes necessary. The project, shown in Figure 9-3, can handle easily at least 5 MΩ. Furthermore, it requires only two resistors, a meter, and a power supply. No values are shown. Don't panic. You are going to use your knowledge of Ohm's law to calculate the values.

The components arrangement is called a *series-ohmmeter circuit.* The meter is in series with the power supply and with R1 and R2. We will assume that R1 is a 3,600-Ω resistor. Your meter may have a different value from the one I used, which was a 50-μA unit with 100-Ω resistance. To determine the value of R2, use the following calculations. What you learn will allow you to substitute any meter you wish and even change the value of R1.

R2 is a variable linear potentiometer. Its purpose is to allow you to adjust the meter circuit to suit the meter you have. Once it is set, you can forget about it as long as your power supply remains constant. The voltage source is any 30-V DC.

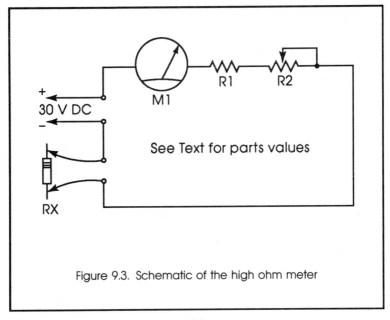

Figure 9.3. Schematic of the high ohm meter

With a supply of 30 volts the total resistance of the series circuit needed to produce 50 μA of current movement can be found in the formula

$$R = V/I$$

where $V = 30$ V and $I = 50$ μA.

$$R = 30 \text{ V} / 0.00005 \text{ A} = 600,000 \ \Omega$$

The meter's resistance is 100 Ω and R1 is 3600 Ω. To find the value of R2 we add the resistance of the meter and of R1 and subtract that sum from the amount given.

The result is $600,000 - 3,700 = 596,300$ Ω. So for our purpose a 600-KΩ potentiometer will be close enough. You can trim it down to the exact value.

The principle of the ohmmeter is that the greater the value of the unknown resistor the less will the needle of the meter be deflected. Let's say that the unknown resistor, R_x causes a deflection halfway across the meter face. From that we would know that only half of the current is flowing through the circuit. Then the total resistance across the circuit, including the known resistor, would be 1,200,000 Ω (30/0.00025). That tells us that the known resistor has a value equal to the difference between 1,200,000 Ω and the 600,000 Ω of the meter circuit, or 600,000 Ω. This gives you the halfway mark on the face of the dial. By using the same formula we locate the quarter mark. It indicates that the value of the resistor used was 1.2 Ω. Depending on the meter you install, you will be able to measure accurately enough for most purposes up to 4 or possibly 5 MΩ.

CONSTRUCTION

The size of the chassis depends on the size of the meter you obtain. The meter should have a glass plate that you can remove. Then, being very careful, slip in a piece of paper or thin cardboard on which you mark the resistances. The needles on meters are extremely delicate so work with care. Then replace the glass plate.

Connect the meter circuit to the power supply by means of binding posts in the back of the metering unit. Observe the polarity of the binding posts. The red post should be connected to the positive side of the supply, which goes to the plus side of the meter. But do not turn on the power until you have a resistor you want to measure across the leads. Those binding posts, which hold the resistor, should be on the front of the chassis. Once you have set R2, it will not have to be readjusted unless you use a different voltage.

When the unit is finished, try a few test resistors to see how close you are to the values you marked. Try to get a 5-MΩ, 1 percent resistor and see whether the meter's indication agrees with your calculations. If the meter shows that you are off a little, then adjust R2 until the position of the needle exactly matches what you expect.

A FINAL WORD

You've built a circuit and you've been very careful. You made no wiring mistakes and the components are exactly what the circuit called for. Yet when you turn on the power you are crushed to see that the circuit is not working quite the way it should. In an audio circuit you hear a hum that covers the sound.

Why?

You call yourself an experimenter, so be one. Start experimenting.

The greatest cause of hum in a delicate circuit such as one that handles audio is the power supply. Is it too near an input circuit? Turn the transformer 90 degrees. That often does the trick. If that doesn't work, try putting a metal shield around the supply. Are your wires from component to component longer than they need to be? A wire a few inches too long can prevent the circuit from behaving. Are you using shielded wire? Are the shields grounded properly? These are only a few suggestions.

Don't hesitate to vary the value of a component slightly,

but not on the power supply. Components may be off, and although the variance may be slight, it may be just enough to throw your circuit off balance.

The reason for experimenting is that you learn as you go along. When you've made a successful change, find the reason why. Even experienced design engineers make changes from what they created on paper in a final circuit. You may even improve a circuit by making some component changes.

When all else fails, breadboard your circuit. Maybe you should have done that in the first place. Then you can vary parts without a lot of desoldering.

GLOSSARY OF ELECTRONICS TERMS

Alternating Current—This current changes direction of its flow 60 times per second. Also called 60 hertz.

Breadboarding—A method of temporarily connecting components in a circuit using a nonconductive punched board. It allows you to visualize how the circuit should be assembled.

Direct Current—A current that flows in only one direction, from a negative to a positive pole.

Etched Board—Components connected by selectively etching away copper from a copper-clad board. This results in metallic paths to which the components are soldered.

Gate—An electronic "door" which responds to coding to open or prevent current flow. Four types of gates exist: OR gates, AND gates, NOR gates, and NAND gates.

Hard Wiring—A term used to describe components linked by means of wire soldered to each.

Integrated Circuit (IC)—A number of capacitors, resistors, and transistors etched together into a single unit.

Ohm's Law—Equations that define the relationships among current, voltage, and resistance in a circuit. Current I is proportional to the voltage, V, and inversely proportional to the resistance, R, of the circuit. The law is expressed in three forms: $V = IR$; $I = V/R$; $R = V/I$.

Photographic Process—A means of producing several identical etched boards by masking the paths and exposing to light.

Power Supply—A group of components usually designed to convert AC to DC as well as filtering the resulting current. Some supplies can be varied over a wide range of voltages, remain constant in spite of current demands, and are impervious to short circuits.

Solid state—A component that operates by controlling the electric or magnetic properties of solids.

Volt Ohm Meter—A popular device for measuring voltages, current, and resistance over several ranges.

Wire Wrapping—Using ultrathin wire and a special wiring tool to wrap wire around the pins of ICs to connect them to each other as well as to the leads of other components. No soldering is necessary.

FOR FURTHER READING

Berlin, H. M., *The 555 Timer with Experiments*. Indianapolis: Howard Sams, 1976.

———. *Design of Op Am Circuits*. Indianapolis: Howard Sams, 1980.

Horowitz, M. *Elementary Electricity and Electronics*. Blue Ridge Summit, Pa.: Tab Books. 1986.

Jung, W. G., *IC Op-amp Cookbook*. Indianapolis: Howard Sams, 1975.

Kuecken, J. A. *How to Measure Anything with Electronic Instruments*. Blue Ridge Summit, Pa.: Tab Books. 1981.

Lenk, J. D. *Handbook of Electronic Test Procedures*. Englewood Cliffs, N.J.: Prentice-Hall, 1982.

Risse, J. A. *Electronic Test Instrument Handbook*. Indianapolis: Howard Sams, 1962.

Shields, J. P. *Practical Power Supply Circuits*. Indianapolis: Howard Sams, 1967.

Traister, R. J. *Beginner's Guide to Reading Schematics*. Blue Ridge Summit, Pa.: Tab Books, 1981.

INDEX

ABOUT THE
AUTHOR

George deLucenay Leon is a free-lance writer who specializes in electronics and other technology subjects. He holds a bachelor of science degree in electrical engineering from Wayne State University. Mr. Leon has written a number of other books, including *AF and RF Signal Generators*, *Casebook of Audio Circuits*, and *The Electricity Story*. He has also contributed articles to such magazines as *Family Computing*, *Communications News*, *Radio Electronics*, and *Popular Electronics*. He lives in Staten Island, New York, with his wife Stish, an interior designer.